Machine Learning

Affine Transformation and Manifold with Python

G.R. Liu

University of Cincinnati, USA

2026

Scientech Publisher, Mason, USA

Published by:

ScienTech Publisher, Mason, USA

Website (for ebooks): https://www.sci-en-tech.com/

ISBN: 978-1-946018-06-9 (paperback)

First Edition

Printed in the United States of America

About the Author

Dr. Gui-Rong Liu received his Ph.D. from Tohoku University, Japan, in 1991. He was a Postdoctoral Fellow at Northwestern University, USA, from 1991 to 1993. He was a Professor at the National University of Singapore until 2010. He is currently a Professor at the Department of Aerospace Engineering and Engineering Mechanics, University of Cincinnati, USA. He was the Founder of the Association for Computational Mechanics (Singapore) (SACM) and served as the President of SACM until 2010. He served as the President of the Asia-Pacific Association for Computational Mechanics (APACM) (2010–2013) and an Executive Council Member of the International Association for Computational Mechanics (IACM) (2005–2010; 2020–2026). He authored a large number of journal papers and books, including two bestsellers, Mesh Free Method: Moving Beyond the Finite Element Method and Smoothed Particle Hydrodynamics: A Meshfree Particle Methods. He is the Editor-in-Chief of the International Journal of Computational Methods and served as an Associate Editor for IPSE and MANO. He is the recipient of numerous awards, including the Singapore Defence Technology Prize, NUS Outstanding University Researcher Award, NUS Best Teacher Award, APACM Computational Mechanics Award, JSME Computational Mechanics Award, ASME Ted Belytschko Applied Mechanics Award, Zienkiewicz Medal from APACM, AJCM Computational Mechanics Award, Humboldt Research Award, SACM Medal from the Association of Computational Mechanics (Singapore), and the Master Engineering Educator Award (UC, CEAS). He has been listed as one among the world's top 1% most influential scientists (Highly Cited Researchers) by Thomson Reuters for a number of years.

Contents

Chapter 1

Introduction and Preliminaries

1.1 Introduction

In *What Is Machine Learning?* [13], we introduced machine learning (ML) at a high level and summarized its fundamental concepts. Building on that foundation, *Machine Learning: Datasets for Classification with Python* [11] and *Machine Learning: Datasets for Regression with Python* [15] focused on the construction, examination, and preprocessing of datasets for supervised learning tasks. We are now prepared to discuss the essential computational mechanisms used to construct ML models.

Throughout this booklet, we assume that the ML problems under consideration are **well-posed** in the Hadamard sense [16], meaning that:

1. The features and labels are related in a continuous manner; small perturbations in the feature space do not cause drastic changes in the labels.
2. The true solution is exist and unique; a given set of feature values corresponds to a unique set of target (label) values.

In addition, we assume that the datasets used to train ML models are of **high quality**, characterized by a sufficient number of samples, adequate coverage of the feature space, and reasonably low noise levels.

> A **good ML model** can always be constructed for a **well-posed problem** using **quality data**.

When data are scarce or noise dominates the true signal, improving the dataset is often more reliable and effective than refining the machine learning methodology itself. Such improvements include thoughtful sampling strategies, careful data acquisition, and appropriate data preprocessing techniques [11, 15].

> **No ML model** can outperform the **data** on which it is trained.

As discussed in Ref. [13], there are two broad mechanisms for constructing ML models:

1. **Flow-path mechanisms**, such as those employed in decision tree–based models.
2. **Parameterization mechanisms**, such as those used in neural networks and most modern ML models.

Given our assumptions of well-posed problems and high-quality data, this booklet focuses primarily on ML models built using **parameterization mechanisms**.

1.2 Acknowledgments

This booklet was prepared with the assistance of AI language models, including Gemini, ChatGPT, DeepSeek, and Qwen, which contributed to drafting text, generating code, and verifying technical details. We also draw upon the author's earlier book on machine learning [9]. The cover image was created using Gemini AI and adapted from an original work by Willy Jackson on Wikimedia Commons, licensed under CC BY-SA 4.0.

1.3 Use of the Book and Codes

Individual readers who purchase this book may contact the author at liugr100@gmail.com to request a complimentary soft copy in Jupyter Notebook format. This version includes executable code (subject to periodic updates) and is available for academic use upon registration.

Conditions of Using Codes in This Book:

The following conditions apply to the use of the book and the codes developed by the author, whether in hard copy, soft copy, or any other medium:

- **User Responsibility:** Users assume all risks associated with the application of any part of the book, codes, or techniques.
- **Code Purpose:** The provided codes are primarily intended for **conceptual demonstration** and may not be optimized for computational efficiency or industrial robustness. Most codes have not undergone third-party testing.
- **Limited Use:** The book and its accompanying codes are for the **personal use** of the individual purchaser. Redistribution or unauthorized copying is strictly prohibited without the author's prior consent.
- **No User Support:** Technical support is not provided for the codes.
- **Citation:** Proper citation and acknowledgment are mandatory when referencing the book, codes, ideas, or techniques in any subsequent work.

These codes rely on various external packages and modules. Their performance is highly dependent on the specific versions of Python and the installed libraries. If a code fails to run as expected, a version mismatch is the most likely cause. You can verify your environment versions directly within a Jupyter Notebook cell. For example, to check the current Python version, you can use:

```
!python --version
!jupyter --version
```

```
Python 3.9.16
Selected Jupyter core packages…
IPython          : 8.10.0
ipykernel        : 6.19.2
ipywidgets       : 7.6.5
jupyter_client   : 6.1.12
jupyter_core     : 5.2.0
jupyter_server   : 1.23.4
jupyterlab       : 3.5.3
nbclient         : 0.5.13
nbconvert        : 6.5.4
nbformat         : 5.7.0
notebook         : 6.1.5
qtconsole        : 5.4.0
traitlets        : 5.7.1
```

We begin by importing the essential Python modules for this booklet. This centralizes our dependencies and keeps subsequent code cells clean.

```
import numpy as np # NUMERICAL & DATA PROCESSING
# The fundamental package for scientific computing with Python.

import matplotlib.pyplot as plt # VISUALIZATION & DATA EXPLORATION
# Used to create static, animated, and interactive visualizations.
```

1.4 Concept of Affine Transformation

This booklet focuses first on the most fundamental method used in ML models: the **affine transformation**.

The affine transformation is utilized in the vast majority of parameterized ML models. The following table highlights the ubiquity of affine transformations across the field of machine learning:

Model	Where Affine Transformation Appears
Perceptron (the earliest ML model)	Threshold classifier
Linear Regression (a basic regressor)	Prediction hyperplane

Model	Where Affine Transformation Appears
Logistic Regression (a basic classifier)	Log-odds function
Neural Networks/MLP (most widely used)	Inside every neuron in every layer
CNNs (image processing, object detection)	Convolution filter weights + bias
RNNs (sequential data, language models)	State transitions, nested neural networks
SVM (convex optimization)	Decision boundary & margin formulation
PCA (dimensionality reduction)	Centering + projection
k-Means (clustering)	Voronoi boundaries
Transformers (LLMs)	Linear projections

This list is far from exhaustive. For the sake of consistency, our discussion will primarily focus on neural network-based models.

An affine transformation consists of two primary operations: **linear transformation** and **translation**. While conceptually simple, it is foundational. When applied to geometric objects, it preserves **affinity**, which ensures:

- **Translation**: Objects are shifted in space.
- **Collinearity**: Straight lines remain straight.
- **Ratio**: The ratio of distances between points on a straight line is preserved.
- **Parallelism**: Parallel line segments remain parallel.

This affinity property is critical for a machine learning model's predictive capabilities. This booklet first establishes the mathematical formulation for affine transformations, followed by a detailed exploration of their processes, properties, and behaviors. We will demonstrate their ability to predict functions within the feature space.

Once the core concept is established, we will introduce the **Affine Transformation Unit (ATU)**. These units can be stacked to form an **Affine Transformation Array (ATA)**, allowing for the flexible increase or decrease of dimensionality within affine subspaces. The ATAs are then lined up to form a multilayer perceptron (MLP), known as universal approximator.

1.5 Concept of Manifold in ML

In the framework of modern geometric learning [21], supervised machine learning can be viewed as a **manifold alignment** problem. Geometrically, every dataset occupies a specific manifold, and the objective of ML is to develop a **model manifold** that accurately fits this **data manifold**.

The second focal point of this booklet is the manifold concept and the methods by which an ML manifold is constructed. We explore how affine transformations, paired with nonlinear activation functions, facilitate this construction. Major topics include:

- **Global manifold formation**: How flat affine mappings, connected by activation functions in a neural network, create complex geometric structures.
- **Manifold dynamics**: The process of folding and unfolding manifolds within Multi-Layer Perceptrons (MLPs).

We confine our discussions in this booklet to **real analysis**. The values of all variables used are real numbers within a real vector space, denoted as $\mathbb{R}$. This space is proven to be closed under arithmetic operations [10, 20] and under functions defined across the entirety of $\mathbb{R}$ [1, 20]. This **closure property** is fundamental, as it ensures the existence of solutions by guaranteeing that standard operations within an ML model will always result in real numbers within $\mathbb{R}$.

While complex numbers may be required for certain ML architectures—particularly when constructing inverse neural networks [6]—they fall outside the scope of this booklet.

Chapter 2

Basic Formulation of Affine Transformation

The use of affine transformations in machine learning can be justified by biological phenomena observed in the behavior of neurotransmitters [9].

2.1 Biological Motivation: Neurotransmitters

Figure 2.1 illustrates the mechanism of neurotransmitter transmission within the synaptic cleft of sensory neurons. The molecules of a neurotransmitter are depicted in a pseudo-colored scanning electron microscope image. In this visualization, a terminal button (green) has been opened to reveal the synaptic vesicles (orange and blue) containing the neurotransmitter.

In an artificial neural network, we utilize the affine transformation (often referred to as the $\mathbf{xw}$ operation) to model the interaction between data points, denoted symbolically as $\mathbf{x}$, and learning parameters, denoted as $\mathbf{w}$. This operation serves as a simplified mathematical proxy for the complex information transformation process occurring via neurotransmitters. Because this transformation is of fundamental importance to modern ML models, the underlying theory and its implications are discussed in great detail throughout this booklet.

2.2 Linear Prediction Function: A Basic Hypothesis

In the general field of computational methods, and specifically within function approximation theory, approximating functions using piecewise linear monomials as basis functions is a widely used technique [10]. In neural network models, the linear prediction function serves as the fundamental hypothesis within each neuron.

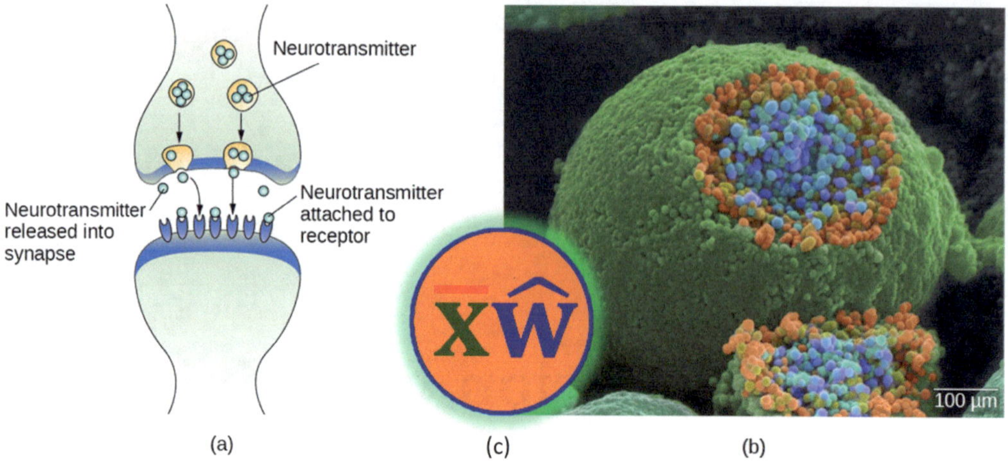

Figure 2.1: (a) Transmission of neurotransmitters; (b) A pseudo-colored image of a neurotransmitter from a scanning electron microscope: a terminal button (green) has been opened to reveal the synaptic vesicles (orange and blue) inside. (c) A typical affine transformation in **xw** form within a neuron of an artificial neural network. (Modified from Psychology 2e by Rose M. Spielman et al., under CC BY 4.0 License. Credit b: modification of work by Tina Carvalho, NIH-NIGMS; scale-bar data from Matt Russell).

Consider a dataset denoted as $\{\mathbf{x}_i, y_i\}_{i=1}^m$. It consists of m data points (or samples), each containing p feature variables defined as a **row vector**:

$$\mathbf{x} = [x_1, x_2, \ldots, x_p] \in \mathbb{R}^p$$

Mimicking the transmission process of neurotransmitters, an artificial neuron introduces $p+1$ learning (or training) parameters. These consist of a single **bias**, denoted as b, and p **weights**, organized as a **column vector**:

$$\mathbf{w} = [w_1, w_2, \ldots, w_p]^\top \in \mathbb{R}^p$$

The most basic hypothesis in ML literature is that the prediction function, often denoted as z, is expressed as follows [9]:

$$z(b, \mathbf{w}; \mathbf{x}) = b + w_1 x_1 + \ldots + w_p x_p = b + \sum_{i=1}^{p} w_i x_i \tag{2.1}$$

Or simply:

$$\boxed{z = b + \mathbf{x}\mathbf{w}} \quad b \in \mathbb{R}, \mathbf{w} \in \mathbb{R}^p, \mathbf{x} \in \mathbb{R}^p \tag{2.2}$$

where $z(\mathbf{w}; \mathbf{x})$ reads as "z is a function of learning parameters b and $\mathbf{w}$ for a given dataset $\mathbf{x}$."

In the foregoing equation, $\mathbb{R}^d$ stands for a real linear vector space of dimension d. This space inherits the same closure property as $\mathbb{R}$, as the difference in dimensionality does not affect the underlying algebraic nature of the field. Therefore, it is proven **closed** under the arithmetic operations used in the equation [10]. This implies that z must reside in $\mathbb{R}$. If the target or label function is linear, it must also reside in $\mathbb{R}$; thus, a solution for the prediction function z is guaranteed to exist within $\mathbb{R}$.

The terminology and physical meaning of these variables are summarized as follows:

- x_i $(i = 1, 2, \ldots, p)$ are feature variables in the feature space $\mathbb{R}^p$. Here, we utilize only *linear basis functions*. Individual data points (or samples) are denoted as $\mathbf{x}_i$ $(i = 1, 2, \ldots, m)$, where m represents the total number of data points.
- $b \in \mathbb{R}$ is known as the **bias** or **intercept**. It is also denoted as $w_0 \in \mathbb{R}$.
- w_i $(i = 1, 2, \ldots, p) \in \mathbb{R}$ are the **weights**, serving as the coefficients for the linear basis functions used in the prediction function.
- Collectively, b and $\mathbf{w}$ are the **learning/training parameters** in each neuron of an ML model. They are not set manually by the modeler; instead, they are iteratively updated during the training process.

The following section will demonstrate that this linear prediction function essentially performs an **affine transformation**.

2.3 Affine Subspaces and Transformation

Equation (2.2) is widely known as the **xw+b form** of affine transformation and is the most common formulation used in machine learning. An affine transformation operates on what is known as an **affine subspace**. Let us now explore the geometric nature of such a subspace.

To do so, we rewrite Eq. (2.2) in the following form:

$$z(\hat{\mathbf{w}}; \mathbf{x}) = b + \mathbf{x}\mathbf{w} = \underbrace{[1 \;\; \mathbf{x}]}_{\overline{\mathbf{x}}} \underbrace{\begin{bmatrix} b \\ \mathbf{w} \end{bmatrix}}_{\hat{\mathbf{w}}} \tag{2.3}$$

Or simply:

$$\boxed{z = \overline{\mathbf{x}}\, \hat{\mathbf{w}}} \tag{2.4}$$

In this form, z is simply the dot product of $\overline{\mathbf{x}}$ and $\hat{\mathbf{w}}$. This is the **xw form**. It is mathematically identical to Eq. (2.2) but expressed in a more concise notation. The augmented vectors used in Eq. (2.4) are defined as follows:

$$\hat{\mathbf{w}} = [b \;\; \mathbf{w}]^\top = [\underbrace{w_0}_{b}, w_1, w_2, \ldots, w_p]^\top \in \mathbb{R}^{p+1} \tag{2.5}$$

where the hat over $\mathbf{w}$ vividly symbolizes the inclusion of the bias term. By "absorbing" the bias, it becomes a new augmented vector containing all the learning parameters in the hypothesis space $\mathbb{R}^{p+1}$. It can also be viewed as a matrix—specifically, one with a single column.

In Eq. (2.4), the row vector $\bar{\mathbf{x}}$ is defined as:

$$\bar{\mathbf{x}} = [1 \;\; \mathbf{x}] = [\underbrace{x_0}_{1}, x_1, x_2, ..., x_p] \in \overline{\mathbb{X}}^p \in \mathbb{R}^{p+1} \tag{2.6}$$

It is important to note that the bar over $\mathbf{x}$ vividly represents the constant 1 prepended to the vector. By combining with this constant, it becomes a new row vector residing in a hyperplane known as the **affine subspace** $\overline{\mathbb{X}}^p$ within $\mathbb{R}^{p+1}$. This is represented by the red plane in Fig. 2.2 for the case where $p = 2$.

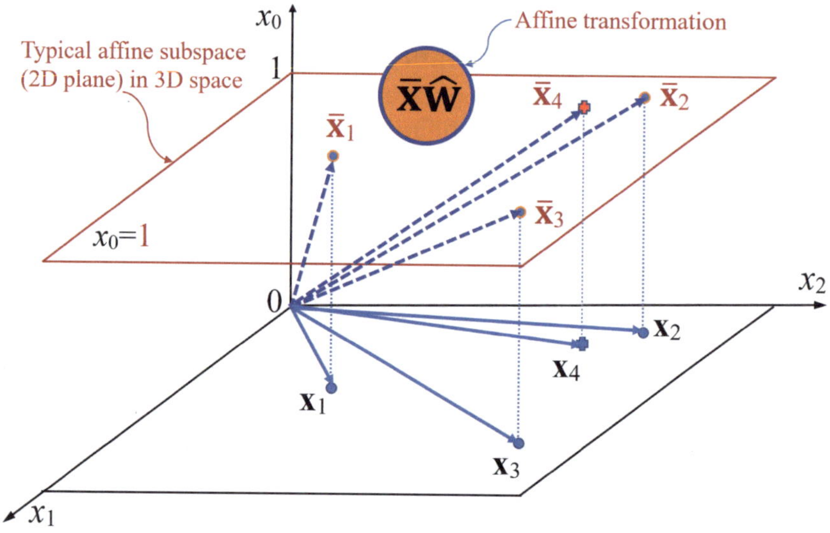

Figure 2.2: Schematic view of an affine subspace where affine transformations are performed on the feature variables $\bar{\mathbf{x}}$ by the learning parameters $\hat{\mathbf{w}}$. Data $\bar{\mathbf{x}}$ and parameters $\hat{\mathbf{w}}$ intersect on this affine subspace.

The affine subspace $\overline{\mathbb{X}}^p$ has a dimension of p. It is not strictly a vector subspace because it does not necessarily pass through the origin. Instead, only the differences between vectors in $\mathbb{R}^{p+1}$ behave like vectors within $\overline{\mathbb{X}}^p$. Because the affine subspace is a p-dimensional plane embedded within a $(p+1)$-dimensional space, it is said to have a **pseudo-dimension** of $p + 1$.

Affine subspaces like the one illustrated in Fig. 2.2 are typical of those used in neural networks. The first dimension of such an affine subspace (the intercept) is always fixed to a unit of 1, as seen in the red plane for $p = 2$. The data points within this affine subspace are denoted as $\overline{\mathbf{x}}_i$ $(i = 1, 2, \ldots, m)$; Fig. 2.2 shows four such points as an example.

In general, a real vector space $\mathbb{R}^d$ contains many possible subspaces. While the intersection value at the first dimension can be any non-zero value, if the intersection is zero, the subspace passes through the origin and becomes a standard vector space. While technically still an affine subspace by definition, this is not the typical form used in neural network models. Transformations occurring in a subspace passing through the origin are purely linear and lack translation, which limits their predictive power. We will explore this further in the following sections.

Furthermore, an affine subspace does not strictly have to align with the x_0 direction orthogonal to the feature space $\mathbb{R}^p$. The one shown in Fig. 2.2 is simply a common example among an infinite number of possible affine subspaces.

Thus, $z(\hat{\mathbf{w}}; \mathbf{x})$ in Eq. (2.2) represents "z as a function of $\hat{\mathbf{w}}$ for a given $\mathbf{x}$" on an **affine subspace**. When the matrix $\mathbf{w}$ acts (from the right) on $\bar{\mathbf{x}}$ as in Eq. (2.2), it produces z. This process is defined as an **affine transformation**. Specifically, Eq. (2.2) performs a transformation from an affine subspace in $\mathbb{R}^3$ to the real vector space $\mathbb{R}^1$.

2.4 From Affine Subspace to Affine Subspace

The affine transformation described by Eq. (2.4) produces a single real number $z \in \mathbb{R}^1$, which belongs to a standard real vector space. However, when constructing Multi-Layer Perceptron (MLP) networks, we must perform affine transformations across multiple hidden layers. This requires the output to also reside within an affine subspace, specifically in $\mathbb{R}^{1+1}$, to maintain the structure necessary for subsequent layers.

This transition can be achieved mathematically by rewriting Eq. (2.4) as:

$$\underbrace{\begin{bmatrix} 1 & z \end{bmatrix}}_{\substack{\overline{\mathbf{z}} \\ \text{in affine subspace} \\ \overline{\mathbb{X}}^1 \in \mathbb{R}^2}} = \underbrace{\begin{bmatrix} 1 & \mathbf{x} \end{bmatrix}}_{\substack{\overline{\mathbf{x}} \\ \text{in affine subspace} \\ \overline{\mathbb{X}}^p \in \mathbb{R}^{p+1}}} \underbrace{\begin{bmatrix} 1 & b \\ \mathbf{0} & \mathbf{w} \end{bmatrix}}_{\substack{\overline{\mathbf{W}} \in \mathbb{R}^{(p+1)\times 2} \\ \text{affine transformation matrix}}} \tag{2.7}$$

or via simple transpose we have:

$$\underbrace{\begin{bmatrix} 1 \\ z \end{bmatrix}}_{\overline{\mathbf{z}}^\top \in \mathbb{R}^2} = \underbrace{\begin{bmatrix} 1 & \mathbf{0}^\top \\ b & \mathbf{w}^\top \end{bmatrix}}_{\overline{\mathbf{W}}^\top \in \mathbb{R}^{2\times(p+1)}} \underbrace{\begin{bmatrix} 1 \\ \mathbf{x}^\top \end{bmatrix}}_{\overline{\mathbf{x}}^\top \in \mathbb{R}^{(p+1)}} \tag{2.8}$$

Equation (2.7) or (2.8) demonstrates that the affine transformation matrix maps a vector from the affine subspace $\overline{\mathbb{X}}^p \subset \mathbb{R}^{p+1}$ to another affine subspace $\overline{\mathbb{X}}^1 \subset \mathbb{R}^2$. This explicitly illustrates how an **affine transformation** is executed within the network architecture.

In Eq. (2.7) or (2.8), we observe the following:

- The affine subspace where the transformation originates is denoted as $\overline{\mathbb{X}}^p$.
- The matrix $\overline{\mathbf{W}}$ (or its transpose $\overline{\mathbf{W}}^\top$) is known as the **affine transformation matrix** [22]. It contains all the learning parameters and is augmented by a unit row (or column) vector. This augmentation ensures the mapping remains consistent from one affine subspace to another.
- The result $\overline{\mathbf{z}}$ resides in a new **affine subspace** of dimension 1, denoted as $\overline{\mathbb{X}}^1$. This is a line within $\mathbb{R}^2$ (since $z \in \mathbb{R}$); thus, $\overline{\mathbb{X}}^1$ has a dimension of 1 and a pseudo-dimension of 2. This subspace is the direct result of the affine transformation matrix acting upon the initial subspace $\overline{\mathbb{X}}^p$.

Because Eq. (2.7) and (2.8) are derived directly from Eq. (2.2) and (2.4), these expressions all represent the same underlying affine transformation. The use of the augmented affine transformation matrix simply ensures the output is properly formatted within a new affine subspace for subsequent processing.

Ultimately, we note that the affine transformation is conceptually identical to the **linear function approximation** hypothesis. The fundamental formulas are exactly the same, merely expressed in a different, more geometrically descriptive form.

2.5 Predictability of Affine Transformations

2.5.1 Role of the Bias: Reproducing Constants

In an affine transformation, the bias $b \in \mathbb{R}$ is a **mandatory component**. This implies that the transformation must occur within an affine subspace with a non-zero intercept. It enables translations, which is a crucial part of the affine transformation. The fundamental reason for this is that if we set $b = 0$ and rely solely on the weights $\mathbf{w}$, the transformation fails to predict even the simplest possible function: a constant. This can be formally demonstrated as follows:

Consider a target (label) function $y(\mathbf{x}) = c$, where c is a constant in $\mathbb{R}$ independent of $\mathbf{x}$. This implies that at the origin, $y(\mathbf{x} = \mathbf{0}) = c$. To successfully predict c using our hypothesis, we must satisfy $z(\mathbf{w}, b; \mathbf{x} = \mathbf{0}) = c$. However, if we omit b from Eq. (2.3), regardless of the values chosen for $\mathbf{w}$, the hypothesis always predicts:

$$z = \mathbf{0} \cdot \mathbf{w} = 0 \tag{2.9}$$

Consequently, the constant $c \in \mathbb{R}$ can never be predicted by the hypothesis without the inclusion of $b \in \mathbb{R}$. A pure linear transformation using only $\mathbf{w}$ is insufficient for robust prediction because it lacks the capacity to model a constant offset.

Conversely, when b is included, we can simply set $b = c$ and $\mathbf{w} = \mathbf{0}$. In this configuration, the affine transformation successfully produces the required constant c.

2.5.2 Role of the Weights: Reproducing a Linear Function

By properly selecting the parameters $\mathbf{w}$ and b in Eq. (2.3), an affine transformation can exactly replicate any given linear target (label) function. This property can be formally demonstrated as follows:

Consider an arbitrary linear function y of independent variables $\mathbf{x} \in \mathbb{R}^p$:

$$y(\mathbf{x}) = \mathbf{x}\mathbf{k} + c \tag{2.10}$$

where $c \in \mathbb{R}$ is a fixed constant and $\mathbf{k} \in \mathbb{R}^p$ is a column vector representing the slope (or gradient) of the function. By simply setting our learning parameters to $\mathbf{w}^* = \mathbf{k}$ and $b^* = c$, our hypothesis becomes:

$$z(\mathbf{w}^*, b^*; \mathbf{x}) = \mathbf{x}\mathbf{k} + c = y(\mathbf{x}) \tag{2.11}$$

Through this specific choice of $\mathbf{w}^*$ and b^*, the target linear function is predicted exactly. This confirms that any arbitrary linear relationship within $\mathbb{R}^p$ can be captured by the affine transformation in Eq. (2.2). In this context, the weights act as the gradient components of the linear model.

This reproducibility property of affine transformations is mathematically analogous to the behavior of linear elements used in the **Finite Element Method (FEM)** [18], where local basis functions are used to approximate a global field.

2.5.3 Affine Transformation in Minimization Procedure

For more complex problems, the underlying label function is typically unknown. Instead, we rely on a dataset [11, 15]. In such cases, we apply an affine transformation coupled with a standard minimization process to determine the learning parameters for both law-based and machine learning methods.

Here, we demonstrate this process to identify the optimal parameters $\mathbf{w}^*$ and b^* for predicting linear functions. Adopting the $\mathbf{xw}$ convention, we rewrite the target linear label function from Eq.(2.10) as:

$$y(\mathbf{x}) = \overline{\mathbf{x}}\,\hat{\mathbf{k}} \tag{2.12}$$

where $\overline{\mathbf{x}}$ represents a data point within the feature space, and $\hat{\mathbf{k}} = [c \;\; \mathbf{k}]^\top \in \mathbb{R}^{p+1}$ is the vector of coefficients.

Step 1: Define a loss function in terms of the learning parameters. The loss function quantifies the discrepancy between the hypothesis in Eq.(2.4) and the label function in Eq.(2.12). While various formulations exist, the most widely used is the L_2 error function, which calculates the squared error:

$$\begin{aligned}\mathcal{L}(z(\hat{\mathbf{w}})) &= \|z(\hat{\mathbf{w}};\overline{\mathbf{x}}) - y(\overline{\mathbf{x}}))\|^2 \\ &= (\overline{\mathbf{x}}\,\hat{\mathbf{w}} - \overline{\mathbf{x}}\,\hat{\mathbf{k}})^\top(\overline{\mathbf{x}}\,\hat{\mathbf{w}} - \overline{\mathbf{x}}\,\hat{\mathbf{k}}) \\ &= (\hat{\mathbf{w}}^\top - \hat{\mathbf{k}}^\top)[\overline{\mathbf{x}}^\top\overline{\mathbf{x}}](\hat{\mathbf{w}} - \hat{\mathbf{k}})\end{aligned} \tag{2.13}$$

The loss function $\mathcal{L}$ is a scalar function of the prediction $z(\hat{\mathbf{w}})$, which in turn depends on the vector of learning parameters $\hat{\mathbf{w}}$. Consequently, $\mathcal{L}(\hat{\mathbf{w}})$ is a functional that is quadratic in $\hat{\mathbf{w}}$, mapping a vector $\hat{\mathbf{w}} \in \mathbb{R}^{p+1}$ to a non-negative real number in $\mathbb{R}$.

In the second line of Eq.(2.13), we apply standard matrix algebra: the transpose is moved into the first set of parentheses, and we then factor out $\overline{\mathbf{x}}^\top$ and $\overline{\mathbf{x}}$ to form the outer-product matrix $[\overline{\mathbf{x}}^\top\overline{\mathbf{x}}]$. This results in a $(p+1) \times (p+1)$ symmetric matrix of rank 1. Since $(\hat{\mathbf{w}} - \hat{\mathbf{k}})$ is a vector, Eq.(2.13) represents a standard quadratic form. If the matrix $[\overline{\mathbf{x}}^\top\overline{\mathbf{x}}]$ were Symmetric Positive Definite (SPD), $\mathcal{L}$ would possess a unique minimum at $(\hat{\mathbf{w}} - \hat{\mathbf{k}}) = 0$, or $\hat{\mathbf{w}}^* = \hat{\mathbf{k}}$. This would prove that the prediction function is **capable** of uniquely reproducing any linear function in the feature space. However, because $[\overline{\mathbf{x}}^\top\overline{\mathbf{x}}]$ has a rank of only 1, further manipulation is required to gain deeper insight.

Step 2: Minimize the scalar loss function with respect to the vector $\hat{\mathbf{w}} \in \mathbb{R}^{p+1}$.

To find the optimal $\hat{\mathbf{w}}$, we locate the stationary point of $\mathcal{L}$ by setting its gradient to zero:

$$\frac{\partial \mathcal{L}(\hat{\mathbf{w}})}{\partial \hat{\mathbf{w}}} = 2\overline{\mathbf{x}}^\top\overline{\mathbf{x}}(\hat{\mathbf{w}} - \hat{\mathbf{k}}) = 0 \tag{2.14}$$

In this derivation, we again utilize the symmetry of $\overline{\mathbf{x}}^\top\overline{\mathbf{x}}$. Equation (2.14) is satisfied when $\hat{\mathbf{w}} = \hat{\mathbf{w}}^* = \hat{\mathbf{k}}$, yielding $\mathbf{w}^* = \mathbf{k}$ and $b^* = c$. This is consistent with previous results and confirms that the prediction function can reproduce any linear function in the feature space.

However, since $\overline{\mathbf{x}}^\top\overline{\mathbf{x}}$ is rank-deficient (rank $= 1$), the solution $\hat{\mathbf{w}}^* = \hat{\mathbf{k}}$ is not unique. There may exist infinitely many solutions where $(\hat{\mathbf{w}} - \hat{\mathbf{k}}) \neq 0$ that still satisfy Eq.(2.14); these solutions reside in the null space of $\overline{\mathbf{x}}^\top\overline{\mathbf{x}}$.

This analysis underscores the necessity of a dataset containing a sufficient number of high-quality data points to generate more equations via Eq.(2.14). In such cases, the vector $\overline{\mathbf{x}}$ is replaced by a data matrix $\overline{\mathbf{X}}$, ensuring that the normal matrix $[\overline{\mathbf{X}}^\top\overline{\mathbf{X}}]$ is SPD. These considerations are central to the **solution existence theory** [8, 14].

Let us examine two examples.

2.5.4 Examples on Predictability

Example I:

Consider a problem characterized by only one feature: x. Suppose we are provided with a single data point pair (x_1, y_1). The normal matrix $\overline{\mathbf{x}}^\top\overline{\mathbf{x}}$ in Eq.(2.14) can be constructed using this discrete data, where the vector $\overline{\mathbf{x}}$ is now denoted as the data matrix $\overline{\mathbf{X}}$:

$$\overline{\mathbf{X}} = \begin{bmatrix} 1 & x_1 \end{bmatrix} \tag{2.15}$$

In this case, the normal matrix $\overline{\mathbf{X}}^\top\overline{\mathbf{X}}$ is a 2×2 matrix:

$$\overline{\mathbf{X}}^\top\overline{\mathbf{X}} = \begin{bmatrix} 1 \\ x_1 \end{bmatrix} \begin{bmatrix} 1 & x_1 \end{bmatrix} = \begin{bmatrix} 1 & x_1 \\ x_1 & x_1^2 \end{bmatrix} \tag{2.16}$$

Note that we are operating within an affine subspace. It is evident that the two columns (or rows) of the matrix $\overline{\mathbf{X}}^\top\overline{\mathbf{X}}$ are linearly dependent (parallel); consequently, this matrix is singular, regardless of the specific values of x_1 and y_1. This singularity confirms that a single data point is insufficient to satisfy the condition $(\hat{\mathbf{w}} - \hat{\mathbf{k}}) = 0$ in Eq.(2.14) for a model with one feature.

Example II:

Considering again a problem with one feature: x. Suppose now we have two distinct data point pairs (x_1, y_1) and (x_2, y_2). Using the data point in the affine subspace, the normal matrix $\overline{\mathbf{x}}^\top\overline{\mathbf{x}}$ in Eq.2.14 can be formed as:

$$\overline{\mathbf{X}}^\top\overline{\mathbf{X}} = \begin{bmatrix} 1 & 1 \\ x_1 & x_2 \end{bmatrix} \begin{bmatrix} 1 & x_1 \\ 1 & x_2 \end{bmatrix} = \begin{bmatrix} 2 & x_1 + x_2 \\ x_1 + x_2 & x_1^2 + x_2^2 \end{bmatrix} \tag{2.17}$$

In this case, the matrix is nonsingular as long as $x_1 \neq x_2$, which can be confirmed using the following snippet:

```
import sympy as sp
x1, x2 = sp.symbols('x1, x2')                          # Define symbolic symbols

XTX = sp.Matrix([[  2,          x1+x2  ],              # Define the normal matrix
                 [x1+x2, x1**2+x2**2]])

XTX_inv = XTX.inv() # Compute inverse
print(f"\nInverse of XTX:")
XTX_inv #.subs(x2,x1)
```

```
Inverse of XTX:
```

$$\begin{bmatrix} \frac{x_1^2+x_2^2}{x_1^2-2x_1x_2+x_2^2} & \frac{-x_1-x_2}{x_1^2-2x_1x_2+x_2^2} \\ \frac{-x_1-x_2}{x_1^2-2x_1x_2+x_2^2} & \frac{2}{x_1^2-2x_1x_2+x_2^2} \end{bmatrix}$$

Because $\overline{\mathbf{X}}^\top \overline{\mathbf{X}}$ in Eq.(2.14) is now nonsingular (for any $x_1 \neq x_2$) and is Symmetric Positive Definite (SPD), the condition $(\hat{\mathbf{w}} - \hat{\mathbf{k}}) = 0$ is guaranteed and unique. Consequently, the linear function defined in Eq.(2.12) is exactly reproduced. It is important to note that while the **existence of a solution** is independent of the label values in the dataset, the specific **value of the solution** itself depends on them.

We have now analytically solved a machine learning problem using a standard minimization procedure to predict a continuous function. Although the example examined is simple, the analysis reveals fundamental issues regarding the predictive capabilities of ML models.

It should be noted that this procedure does not automatically guarantee that such a solution can be found for general problems where the true underlying function is unknown (unlike $\hat{\mathbf{k}}$ in this example). For more complex problems involving discrete datasets, we typically rely on computational methods to solve minimization problems with non-trivial loss functions. Nevertheless, the core concept remains unchanged: the solution exists if the data is sufficient and of good quality, and the primary challenge lies in the methodology used to locate it.

Chapter 3

Affine Transformation Unit (ATU)

3.1 $p \to 1$ net

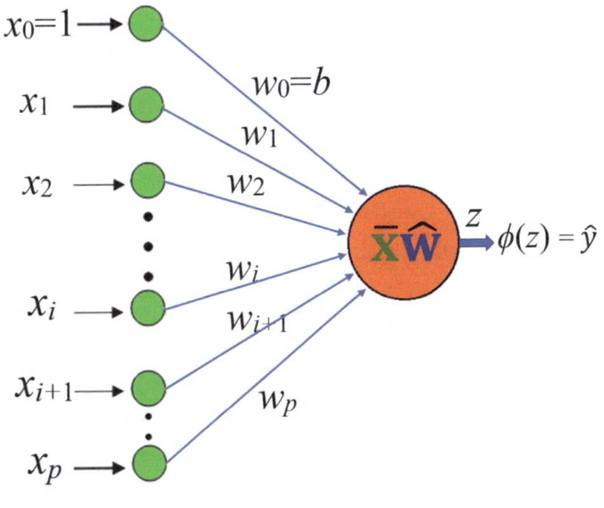

Figure 3.1: A $p \to 1$ neural network featuring an input layer that accommodates p features on an affine subspace (effectively $p+1$ inputs). With a single neuron at the output, it produces $z \in \mathbb{R}^1$, providing the capacity to predict any linear function in the feature space. Furthermore, when z is passed through a nonlinear activation function ϕ, the output becomes $\hat{y} = \phi(z)$, enabling the model to predict nonlinear functions.

The affine transformation defined in Eq. (2.4) can be used to construct a $p \to 1$ neural network, which serves as a basic neural unit for building more powerful neural network architectures. This unit is capable of representing arbitrary linear functions within a given feature space. Typical architectures of a $p \to 1$ network are illustrated in Fig. (3.1).

This $p \to 1$ network constitutes an **Affine Transformation Unit (ATU)**. It is capable of predicting arbitrary linear functions $z(\mathbf{x})$ of the feature variables $\mathbf{x} = [x_1, x_2, \dots, x_p]$ using a set of learning parameters $\mathbf{w}_i$, where w_0 corresponds to the bias b. For a fixed set of learning parameters, the unit produces a scalar output $z(\mathbf{x}_j) \in \mathbb{R}^1$ for any given data point (sample) $\mathbf{x}_j$.

The ATU serves as a fundamental building block of neural networks; when stacked sequentially or arranged in parallel, multiple ATUs can be combined to form complex deep neural network architectures.

3.2 Predictability of the ATU

Consider a high-quality dataset for regression tasks. The true label value y_j corresponding to a sample $\mathbf{x}_j$ must lie in $\mathbb{R}^1$.

Since all label values reside in $\mathbb{R}$, a solution to a well-posed regression problem necessarily exists. If a machine learning (ML) model is capable of producing real-valued outputs spanning $\mathbb{R}$ by tuning its learning parameters through operations that preserve the **closure property** of $\mathbb{R}$, the model is said to possess *predictability*.

- **Linear Dependence:** If y_j depends linearly on the feature variable $\mathbf{x}$, the network can produce a prediction $\hat{y}_j$ that closely approximates the true label y_j, provided that an optimal set of learning parameters is identified. This capability follows directly from the **closure property** of the real vector space $\mathbb{R}^1$ and the fact that affine transformations involve only arithmetic operations. Identifying such parameters requires an appropriate training strategy and a sufficient number of high-quality samples.

- **Nonlinear Dependence:** If y_j depends nonlinearly on the feature variable $\mathbf{x}$, a **nonlinear function** (activation function), denoted by $\phi(\cdot)$, is introduced to wrap the affine output z. In most ML models, activation functions are defined over the entire real line $\mathbb{R}$. Owing to the **closure property** of $\mathbb{R}^1$, the resulting network can still be made capable of producing $\hat{y}_j \approx y_j$. A canonical example is logistic regression, which employs the logistic activation function.

- **Feature Transformation:** Due to the nonlinearity of $\phi(\cdot)$, the transformed output $\phi(z)$ becomes linearly independent of the original feature variables. It therefore acts as a new feature variable for subsequent neurons.

- **Deep Networks:** When $y_j(\mathbf{x})$ exhibits highly nonlinear behavior, multiple ATUs can be connected in sequence (horizontally), with the output $\phi(z)$ of one unit serving as the input to the next layer. Such a configuration constitutes a deep neural network.

As shown in Fig. (3.1), the affine transformation is performed within each neuron. After the affine transformation and the application of a nonlinear activation, the ATU establishes a global mapping from the feature space $\mathbb{R}^p$ to the label space $\mathbb{R}^1$. The total number of learning parameters (the P-number) for a single ATU is $(p+1)$.

3.3 Existence of Unique Solution

Although a solution exists in $\mathbb{R}$ and the ML model is theoretically capable of reaching it, the uniqueness of the solution must also be ensured. In the absence of uniqueness, the model may admit multiple parameter sets that satisfy the training criteria. This issue is resolved by imposing appropriate conditions on the dataset. The fundamental requirement is:

> The number of quality samples in the dataset, m, must be greater than $(p+1)$, which is the total pseudo-dimension of the ATU.

This condition corresponds to the **Neurons–Samples Theorem** (Chapter 13 of Ref. [9], Ref. [7]) applied to a single ATU. Equivalently, it can be stated as:

> The number of quality samples m must exceed the total number of neurons $(p+1)$ in the ATU.

Here, the total number of neurons includes those in the input layer (the green neurons in Fig. 3.1). The output neuron (red) is excluded from this count, as no further affine transformation is applied after it.

Key Conclusions:

1. An ATU is capable of predicting any linear function in the feature space.
2. ATUs wrapped with nonlinear activation functions and connected in sequence are capable of predicting any continuous nonlinear function.
3. An ATU can produce a unique solution that approximates the true solution in the dataset if $m \geq (p+1)$.

Chapter 4

Affine Transformation Array (ATA)

4.1 Affine Transformation in Higher Dimensions

To demonstrate how Eq.(2.4) is explicitly utilized to perform an affine transformation between higher-dimensional affine subspaces, we apply the following maneuver using matrix formulations:

4.1.1 Affine Subspace to Vector Space

First, by utilizing each vector $\hat{\mathbf{w}}_i \in \mathbb{R}^{p+1}$ (for $i = 1, 2, \dots, k$) and applying Eq.(2.4), we obtain a mapping from the input affine subspace to a vector space.

$$z_i = \overline{\mathbf{x}}\,\hat{\mathbf{w}}_i \tag{4.1}$$

Next, we form the following vector:

$$\begin{aligned}\mathbf{z} = [z_1, z_2, ..., z_k] &= \left[\overline{\mathbf{x}}\begin{bmatrix} b_1 \\ \mathbf{w}_1 \end{bmatrix}, \overline{\mathbf{x}}\begin{bmatrix} b_2 \\ \mathbf{w}_2 \end{bmatrix}, ..., \overline{\mathbf{x}}\begin{bmatrix} b_p \\ \mathbf{w}_k \end{bmatrix}\right] \\ &= \overline{\mathbf{x}}\begin{bmatrix} \mathbf{b} \\ \mathbf{W} \end{bmatrix} = [1 \quad \mathbf{x}]\begin{bmatrix} \mathbf{b} \\ \mathbf{W} \end{bmatrix} = \overline{\mathbf{x}}\underbrace{\begin{bmatrix} \mathbf{W}_0 \\ \mathbf{W} \end{bmatrix}}_{\hat{\mathbf{W}}}\end{aligned} \tag{4.2}$$

Or in matrix form:

$$\mathbf{z} = \mathbf{b} + \mathbf{x}\mathbf{W} = \overline{\mathbf{x}}\,\hat{\mathbf{W}} \tag{4.3}$$

where $\mathbf{b} = [b_1, b_2, ..., b_k]$, $\mathbf{W}_0 = [\underbrace{w_{01}}_{b_1}, \underbrace{w_{02}}_{b_2}, ..., \underbrace{w_{0k}}_{b_k}] = \mathbf{b}$, and $\mathbf{W} = [\mathbf{w}_1, \mathbf{w}_2, ..., \mathbf{w}_k]$.

Notice that $\hat{\mathbf{W}}$ is now in $\mathbb{R}^{(p+1)\times k}$. It absorbs all bias terms $\mathbf{b}$ as the zeroth-indexed weights w_0, thus becoming the "augmented" version (denoted by the hat) of $\mathbf{W}$. This matrix collects all the learning parameters required to predict $\mathbf{z} \in \mathbb{R}^k$ by transforming $\overline{\mathbf{x}} \in \mathbb{R}^{p+1}$. Our notation $\hat{\mathbf{W}}$ facilitates the easy tracking of variables. This formulation is frequently used for prediction at the output layer of a neural network, such as the Multilayer Perceptron (MLP), because $\mathbf{z} = \overline{\mathbf{x}}\hat{\mathbf{W}}$ implies that a subsequent affine transformation is no longer necessary at the final stage. We utilize this normal transformation to obtain a resulting vector in the vector space $\mathbb{R}^k$.

4.1.2 Transformation from Affine Subspace to Affine Subspace

To explicitly demonstrate the affine transformation occurring between two affine subspaces, we can further construct the following matrix operation:

$$\underbrace{\begin{bmatrix} 1 \\ \mathbf{z}^\top \end{bmatrix}}_{\overline{\mathbf{z}}^\top} = \underbrace{\begin{bmatrix} 1 & \mathbf{0}^\top \\ \mathbf{b}^\top & \mathbf{W}^\top \end{bmatrix}}_{\overline{\mathbf{W}}^\top} \underbrace{\begin{bmatrix} 1 \\ \mathbf{x}^\top \end{bmatrix}}_{\overline{\mathbf{x}}^\top} = \underbrace{\begin{bmatrix} 1 & \mathbf{0}^\top \\ \mathbf{W}_0^\top & \mathbf{W}^\top \end{bmatrix}}_{\overline{\mathbf{W}}^\top} \underbrace{\begin{bmatrix} 1 \\ \mathbf{x}^\top \end{bmatrix}}_{\overline{\mathbf{x}}^\top} \tag{4.4}$$

or simply

$$\overline{\mathbf{z}}^\top = \overline{\mathbf{W}}^\top \overline{\mathbf{x}}^\top \tag{4.5}$$

In its transposed form:

$$\underbrace{\begin{bmatrix} 1 & \mathbf{z} \end{bmatrix}}_{\overline{\mathbf{z}} \in \mathbb{R}^{(k+1)}} = \underbrace{\begin{bmatrix} 1 & \mathbf{x} \end{bmatrix}}_{\overline{\mathbf{x}} \in \mathbb{R}^{(p+1)}} \underbrace{\begin{bmatrix} 1 & \mathbf{W}_0 \\ \mathbf{0} & \mathbf{W} \end{bmatrix}}_{\overline{\mathbf{W}} \in \mathbb{R}^{(p+1)\times(k+1)}} \tag{4.6}$$

or simply

$$\overline{\mathbf{z}} = \overline{\mathbf{x}}\,\overline{\mathbf{W}} \tag{4.7}$$

where $\mathbf{0} = [0, 0, \dots, 0]^\top$ is a vector containing p zeros. The matrix $\overline{\mathbf{W}}$ (or its transpose $\overline{\mathbf{W}}^\top$) serves as the affine transformation matrix mapping from the affine subspace $\overline{\mathbb{X}}^p \subset \mathbb{R}^{p+1}$ to the affine subspace $\overline{\mathbb{X}}^k \subset \mathbb{R}^{k+1}$. In this framework, the affine matrix $\overline{\mathbf{W}}$ acts on a row vector from the right, while $\overline{\mathbf{W}}^\top$ acts on a column vector from the left. This operation performs affine transformations across the affine subspaces, taking a given $\overline{\mathbf{x}} \in \overline{\mathbb{X}}^p$ and producing a corresponding $\overline{\mathbf{z}} \in \overline{\mathbb{X}}^k$. This structure is essential for all hidden layers in a Multilayer Perceptron (MLP), as transformations must occur within these affine subspaces to ensure reproducibility across successive hidden layers.

It is important to note that $\overline{\mathbf{W}}$ can be expressed as $[\mathbf{e}_1 \quad \hat{\mathbf{W}}]$, where $\mathbf{e}_1 = [1 \quad \mathbf{0}]^\top$ is the first basis vector of the space $\mathbb{R}^{p+1}$. Because $\overline{\mathbf{W}}$ contains this constant unit vector as its first column, the actual trainable parameters are localized within $\hat{\mathbf{W}}$. In the output layer of an MLP, we transition to using only $\hat{\mathbf{W}}$, as the requirement for maintaining the affine subspace structure concludes at the final prediction.

The formulation provided above explicitly reveals the derivation of the affine transformation weight matrix and demonstrates how traditional weights and biases are integrated into the formal affine transformation matrix.

4.2 Affine Transformations in an Array

The affine transformation defined in Eq. (4.2) can be realized by vertically stacking k individual ATUs, as illustrated in Fig. (4.1). In this configuration, the neurons are *densely connected* (also referred to as *fully connected*), meaning that the output of every neuron in the preceding layer serves as an input to every neuron in the subsequent layer. This architecture forms a $p \to k$ network that maps a feature space in $\mathbb{R}^p$ to a target space in $\mathbb{R}^k$.

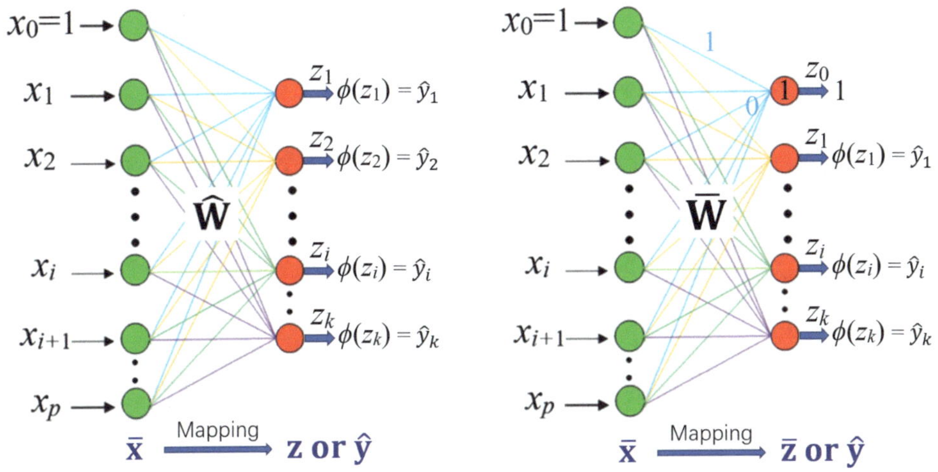

Figure 4.1: A $p \to k$ neural network consisting of an input layer with $(p+1)$ neurons and an output layer of k neurons, producing k distinct prediction functions z_i $(i = 1, 2, \ldots, k)$. Each neuron in the input layer is connected to every neuron in the output layer through its own weight. This stack of ATUs forms an Affine Transformation Array (ATA). Consequently, the $p \to k$ network possesses predictability for k functions in a p-dimensional feature space. Left: mapping $\bar{\mathbf{x}}$ to $\mathbf{z}$; Right: mapping $\bar{\mathbf{x}}$ to the augmented vector $\bar{\mathbf{z}}$.

This ATA structure represents the fundamental linear transformation performed within any neural network layer. By executing these affine operations in parallel, the network can simultaneously learn multiple facets of the input data. When $\bar{\mathbf{x}}$ is mapped to $\bar{\mathbf{z}}$, the affine property is preserved, which enables the recursive application of this construction throughout deep network architectures.

The prediction functions $\mathbf{z}$ produced by an ATA can be expressed as

$$\mathbf{z} = \mathbf{x}\mathbf{W} + \mathbf{b} = \overline{\mathbf{x}}, \hat{\mathbf{W}}, \quad (4.8)$$

where $\mathbf{z}$ is a vector of prediction functions defined as:

- $\mathbf{z} = [z_1, z_2, \ldots, z_k]$, a vector of functions;
- $\hat{\mathbf{W}} = [\mathbf{b} \;\; \mathbf{W}]^\top$, a $(p+1) \times k$ matrix containing all learning parameters for the neurons in the output layer;
- $\mathbf{W} = [w_{ij}]$, $(i = 1, 2, \ldots, p; , j = 1, 2, \ldots, k)$, the weight matrix (part of the learning parameters); and
- $\mathbf{b} = [b_1, b_2, \ldots, b_k]$, the bias vector (part of the learning parameters).

The complete set of learning parameters of a $p \to k$ ATA can be written as a single vector $\hat{\mathbf{w}}$:

$$\hat{\mathbf{w}} = [\hat{\mathbf{w}}_1^\top, \hat{\mathbf{w}}_2^\top, \ldots, \hat{\mathbf{w}}_k^\top]^\top$$

which can also be interpreted as the flattened form of $\hat{\mathbf{W}}$. The total number of learning parameters is therefore

$$P = (p+1) \times k$$

It is evident that the hypothesis space expands rapidly as the number of ATUs in an ATA increases. Equation (4.8) provides a compact matrix representation of this collection of concurrent affine transformations.

Crucially, each output $z_j(\mathbf{w}_j, b_j)$ is computed via Eq. (2.4) using its own dedicated set of weights $\mathbf{w}_j$ and bias b_j. This independence ensures that each z_j, for $j = 1, 2, \ldots, k$, can represent a distinct functional relationship. Consequently, the ATA shown in Fig. 4.1 establishes the most fundamental mapping suitable for k-dimensional regression problems derived from a dataset with p features. Furthermore, in the special case where $k = p$, the ATA performs a $p \to p$ affine transformation within the same affine subspace, enabling operations such as rotation, scaling, and translation of the feature coordinates.

4.3 Special Cases of Affine Transformations

In machine learning models, learning parameters are optimized through training. By examining specific parameter configurations in Eq. (4.5), we gain insight into how an affine transformation manipulates the data manifold within an affine subspace and how these transformations are progressively discovered through backpropagation in an **Affine Transformation Array (ATA)**.

4.3.1 Special Cases of Learning Parameter Setting

Several important affine transformations arise as special cases, each providing geometric intuition for the role of different learning parameters.

Case 1: Zero Initialization If all learning parameters are set to zero, i.e., $\hat{\mathbf{W}} = \mathbf{0}$, then the transformation reduces to $[1 \;\; \mathbf{z}] = [1 \;\; \mathbf{0}]$. In this case, every data point $[1 \;\; \mathbf{x}]$ in the affine subspace collapses to a single point. This degeneracy illustrates why zero initialization is ineffective: no meaningful affine transformation can occur, and gradients may vanish or become identical across neurons. Consequently, weights are typically initialized with small random values drawn from carefully designed distributions (e.g., Xavier or He initialization [4]) to enable learning.

Case 2: The Identity Transformation If the bias vector is set to $\mathbf{b} = \mathbf{0}$ and the weight matrix to $\mathbf{W} = \mathbf{I}$, the transformation becomes $[1 \;\; \mathbf{z}] = [1 \;\; \mathbf{x}]$. Each point in the affine subspace remains unchanged. This identity mapping plays a central role in modern architectures such as Residual Networks (ResNets) [3], where skip connections allow information to pass through layers without distortion.

Case 3: Pure Translation If the bias is set to a constant vector $\mathbf{b} = \mathbf{c}$ and the weight matrix remains $\mathbf{W} = \mathbf{I}$, then $[1 \;\; \mathbf{z}] = [1 \;\; \mathbf{c} + \mathbf{x}]$. All points in the affine subspace are translated by the vector $\mathbf{c}$. This case highlights the geometric role of the bias term, which shifts the entire feature space relative to the origin.

Case 4: Partial Linear Prediction Consider the configuration $\mathbf{b} = [c, 0, \dots, 0]^\top$ and $\mathbf{W} = [\mathbf{k}, \mathbf{e}_2, \dots, \mathbf{e}_p]$, where $\mathbf{e}_i$ denotes the standard basis vectors of $\mathbb{R}^p$. In this case, the components $z_i = x_i$ for $(i = 2, \dots, p)$ remain unchanged, while the first output is given by

$$z_1 = \mathbf{x}\mathbf{k} + c \tag{4.9}$$

This configuration aligns with Eq. (2.11) and demonstrates that predicting a linear function in the feature space is mathematically equivalent to performing an affine transformation in the augmented affine subspace. Geometrically, the vector $\mathbf{k}$ represents the gradient of the linear function and governs the rotation of the prediction hyperplane.

4.3.2 Learning Dynamics during Backpropagation

Within an ATA, backpropagation serves as the mechanism that guides the learning parameters from an initial configuration—typically near Case 1—toward a solution that effectively combines Cases 2, 3, and 4.

Breaking Symmetry (Beyond Case 1) A zero or near-zero initialization corresponds to a collapsed affine subspace. During early backpropagation steps, the gradient $\frac{\partial \mathcal{L}}{\partial \mathbf{w}}$ introduces small asymmetric updates that break this symmetry. Without such differentiation, all neurons would receive identical gradients and fail to learn distinct features. Moving away from this collapsed state allows the ATA to expand its hypothesis space.

Identity Bias and Residual Learning (Case 2) In deep multilayer perceptrons, directly learning the identity transformation can be difficult. Residual connections explicitly define the transformation as $\mathbf{z} = \mathbf{x} + \mathcal{F}(\mathbf{x})$, effectively embedding Case 2 as the default mapping. Backpropagation then focuses on learning only the residual function $\mathcal{F}(\mathbf{x})$, significantly improving optimization in deep networks.

Bias Adaptation and Data Centering (Case 3) During training, the bias vector $\mathbf{b}$ often adapts rapidly. From a geometric perspective, this corresponds to translating the affine subspace to better align with the data distribution. If the target labels have a nonzero mean, backpropagation naturally steers the parameters toward a translation that reduces the L_2 loss by centering the predictions.

Gradient-Based Rotation (Case 4) The most computationally intensive aspect of learning involves adjusting the weight matrix $\mathbf{W}$. The gradient $\frac{\partial \mathcal{L}}{\partial \mathbf{W}}$ determines how the prediction hyperplanes must rotate in the p-dimensional feature space to minimize the discrepancy between the predictions $\mathbf{z}$ (or $\phi(\mathbf{z})$) and the true labels $\mathbf{y}$.

4.3.3 An Intuitive Learning Trajectory

The minimization procedure described in Section 2.5 can be viewed as a progression through these special cases:

1. **Initialization:** Small random weights (near Case 1).
2. **Translation:** Rapid bias adjustment to match the data mean (Case 3).
3. **Rotation:** Iterative refinement of $\mathbf{W}$ to align prediction slopes with data trends (Case 4).
4. **Refinement:** Fine-tuning toward a unique optimal solution $\hat{\mathbf{w}}^* = \hat{\mathbf{k}}$.

This perspective provides a unified geometric interpretation of affine transformations and clarifies how backpropagation systematically constructs increasingly expressive mappings from simple foundational cases.

4.4 Predictability of ATA and MLP

Consider a simple regression dataset denoted by $\{\mathbf{x}_i, \mathbf{y}_i\}_{i=1}^{m}$. It consists of m data points (samples), where each sample has p feature variables represented by a row vector

$$\mathbf{x} = [x_1, x_2, \dots, x_p] \in \mathbb{R}^p \tag{4.10}$$

and k label variables represented by a column vector

$$\mathbf{y} = [y_1, y_2, \dots, y_k] \in \mathbb{R}^k \tag{4.11}$$

This vector-valued function constitutes the prediction target of the learning problem.

4.4.1 Model Predictability

Since the label variables take values in $\mathbb{R}$, a solution to any well-posed regression problem necessarily exists within that space. A machine learning (ML) model is said to possess **predictability** if it can generate real-valued outputs spanning $\mathbb{R}$ by tuning its learning parameters—also defined in $\mathbb{R}$—using operations that satisfy the **closure property** of the real number system.

- **Linear function predictability:** If the components of $\mathbf{y}_i$ depend linearly on the feature variables $\mathbf{x}$, the model can produce an output $\mathbf{z}_i$ that closely approximates the true label vector $\mathbf{y}_i$ (under an appropriate metric). This follows directly from the **closure property** of real vector spaces $\mathbb{R}^d$ and the fact that affine transformations involve only arithmetic operations. Although an optimal set of learning parameters exists, identifying it requires an effective training strategy and a sufficient number of high-quality samples.
- **Nonlinear function predictability:** If $\mathbf{y}_i$ depends nonlinearly—but continuously—on the feature variables $\mathbf{x}$, a **nonlinear activation function** $\phi(\cdot)$ is introduced. This function is applied element-wise to the affine output $\mathbf{z}$ to produce the prediction $\hat{\mathbf{y}}_i$. Owing to the **closure property** of real vector spaces and the fact that activation functions are defined over all of $\mathbb{R}$, the resulting network can still approximate $\mathbf{y}_i$ arbitrarily well, provided suitable learning parameters are chosen.
- **Highly nonlinear function predictability:** Because $\phi(\cdot)$ is nonlinear, the transformed output $\phi(\mathbf{z})$ becomes linearly independent of the original feature variables (or the outputs of the preceding layer). It therefore serves as a new set of feature variables for subsequent layers. Consequently, when $\mathbf{y}_i(\mathbf{x})$ exhibits highly nonlinear behavior, multiple ATAs can be connected sequentially (horizontally), using $\phi(\mathbf{z})$ as the input features at each stage. This construction yields a deep Multilayer Perceptron (MLP), as shown in Fig. (4.2), which is capable of approximating any continuous, highly nonlinear function.

As illustrated in Fig. (4.2), a large number of affine transformations are performed—one within each gray neuron. Collectively, these affine mappings and nonlinear activations enable the MLP to form a global mapping from the feature space $\mathbb{R}^p$ to the label space $\mathbb{R}^k$, regardless of the relative dimensions of p and k (including the case $p < k$). Autoencoders [9] and trumpet networks [6] are representative examples of such architectures. In this sense, an MLP can be viewed as a **giant affine transformation machine, MLP**. The total number of learning parameters (the P-number) may be extremely large; a systematic formulation for computing it is provided in Chapter 13 of Ref. [9].

4.4.2 Existence of a Unique Solution

Although a well-posed problem admits a true solution in $\mathbb{R}$ and an ML model is capable of approximating it, uniqueness of the solution must also be ensured. Without appropriate constraints, the model may converge to multiple valid solutions, as discussed in Section 2.5. This issue is resolved by imposing sufficient requirements on the dataset. The fundamental condition is:

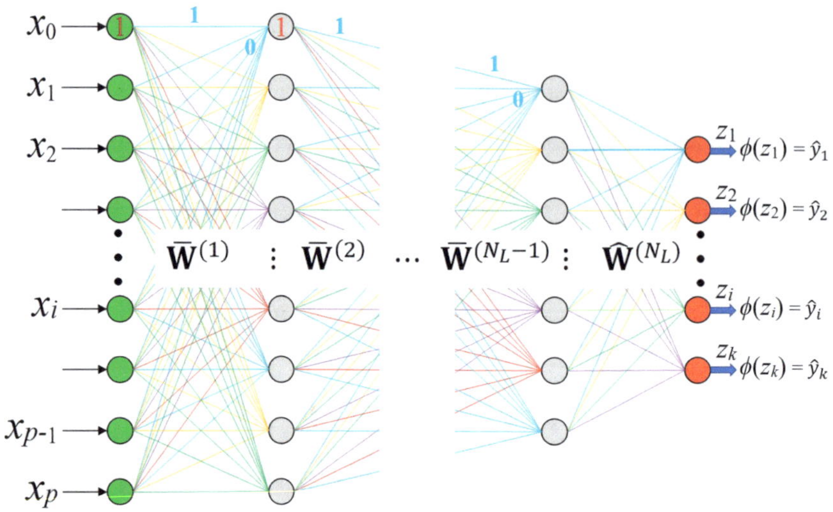

Figure 4.2: A giant affine transformation machine: An MLP with an input layer of $(p+1)$ neurons (green), an output layer of k neurons (red), and multiple hidden layers composed of numerous neurons (gray). The network performs a $p \to k$ mapping and is capable of predicting k continuous label functions in a p-dimensional feature space.

> The number of quality samples in the dataset, m, must be larger than P^*, where P^* denotes the total pseudo-dimension of all affine subspaces associated with all ATAs in the MLP.

This result follows from the **Neurons–Samples Theorem** (Chapter 13 of Ref. [9], Ref. [7]) and can be simplified as:

> The number of quality samples m must exceed the total number of neurons in the MLP, denoted by P^*.

The neuron count includes all neurons in the input layer (green) and the hidden layers (gray). Neurons in the output layer are excluded, as no further affine transformation is applied after them.

For a single ATU, the P^*-number is $(p + 1)$. More generally, for an MLP, the P^*-number is typically much smaller than the total number of learning parameters (the P-number). For an MLP with uniform hidden-layer width h, the relationship is

$$P^*\text{-number} = \frac{P\text{-number}}{h} \tag{4.12}$$

For example, if $h = 100$, the P^*-number is 100 times smaller than the P-number.

In summary:

An ATA is capable of predicting any linear function in high-dimensional feature spaces. When ATAs are linked and wrapped with nonlinear activation functions, they form an MLP capable of predicting any continuous nonlinear function in high dimensions—this is the essence of the **universal approximation theorem**.

An MLP can yield a unique solution that closely approximates the true solution of a well-posed problem, provided that

$$\boxed{m \geq P^*} \tag{4.13}$$

Chapter 5

Properties of Affine Transformation

The study of the properties of affine transformations is essential for understanding how objects within the feature space of a dataset are transformed as they pass through the layers of a neural network.

5.1 Mapping within the Same Affine Subspace

To simplify the discussion, we first examine affine mappings within a single affine subspace. In this case, we set $k = p$ in Eq.(4.6), meaning the affine transformation maps from the affine subspace $\overline{\mathbb{X}}^p$ back to $\overline{\mathbb{X}}^p$. Consequently, the affine transformation matrix $\overline{\mathbf{W}}$ in Eq.(4.7) is square. For this discussion, we assume by default that $\overline{\mathbf{W}}$ is invertible; exceptional cases will be addressed separately. Such affine transformations possess several key properties.

5.1.1 Injective (One-to-One)

If $\overline{\mathbf{W}}$ is invertible, the affine transformation is **injective**, meaning the mapping is **one-to-one**: distinct input points are always mapped to distinct output points.

Proof:

Consider two distinct points $\mathbf{x}_1$ and $\mathbf{x}_2$ in the feature space such that $\mathbf{x}_1 \neq \mathbf{x}_2$. Applying Eq.(4.6) to both, we have:

$$\begin{aligned} \overline{\mathbf{z}}(\mathbf{x}_1) &= \overline{\mathbf{x}}_1 \, \overline{\mathbf{W}} \\ \overline{\mathbf{z}}(\mathbf{x}_2) &= \overline{\mathbf{x}}_2 \, \overline{\mathbf{W}} \end{aligned} \tag{5.1}$$

The difference between the transformed points is:

$$\overline{\mathbf{z}}(\mathbf{x}_2) - \overline{\mathbf{z}}(\mathbf{x}_1) = (\overline{\mathbf{x}}_2 - \overline{\mathbf{x}}_1)\,\overline{\mathbf{W}} \tag{5.2}$$

Since $\overline{\mathbf{W}}$ is invertible and thus has full rank, the equality $\overline{\mathbf{z}}(\mathbf{x}_2) = \overline{\mathbf{z}}(\mathbf{x}_1)$ can only occur if $\overline{\mathbf{x}}_2 = \overline{\mathbf{x}}_1$ (see Ref. [14]). Therefore, if $\mathbf{x}_1$ and $\mathbf{x}_2$ are distinct, their transformations $\overline{\mathbf{z}}(\mathbf{x}_1)$ and $\overline{\mathbf{z}}(\mathbf{x}_2)$ must also be distinct. This confirms the mapping is injective.

In contrast, if $\overline{\mathbf{W}}$ is rank-deficient, the mapping is no longer injective. In such cases, distinct points (vectors) belonging to the **null-space** of $\overline{\mathbf{W}}$ are all mapped to the zero vector, leading to degenerated geometry.

5.1.2 Inverse Mapping

When $q = p$, the affine transformation matrix is square. If the weight matrix $\mathbf{W}$ is invertible, the augmented matrix $\overline{\mathbf{W}}$ is also invertible and follows this block-matrix formulation:

$$\overline{\mathbf{W}}^{-1} = \begin{bmatrix} 1 & -\mathbf{W}_0\mathbf{W}^{-1} \\ \mathbf{0} & \mathbf{W}^{-1} \end{bmatrix} \quad \text{or} \quad \overline{\mathbf{W}}^{-T} = \begin{bmatrix} 1 & \mathbf{0}^T \\ -\mathbf{W}^{-T}\mathbf{W}_0 & \mathbf{W}^{-T} \end{bmatrix} \tag{5.3}$$

This relationship can be verified by checking that $\overline{\mathbf{W}}\,\overline{\mathbf{W}}^{-1} = \mathbf{I}$ or $\overline{\mathbf{W}}^T\overline{\mathbf{W}}^{-T} = \mathbf{I}$. Under these conditions, the affine transformation is reversible and **bijective**, ensuring a unique one-to-one correspondence for both forward and inverse mappings.

Conversely, if $\overline{\mathbf{W}}$ is rank-deficient, it possesses a non-trivial null-space, and the forward mapping is no longer injective. This leads to **geometric degeneracy**—where distinct features are collapsed onto the same point or a lower-dimensional subspace. Essentially, the network "forgets" specific differences between input features. In such cases, a direct inverse mapping is impossible using the existing learning parameters, and specialized techniques (such as regularization) are required to manage the resulting ill-posed problem [16].

It is important to note that learning parameters are updated dynamically during the training process. If the MLP determines—based on the loss function—that such "forgetting" is beneficial for reducing the error, it may purposely drive $\overline{\mathbf{W}}$ toward a rank-deficient state. This behavior is often observed in dimensionality reduction tasks or feature extraction, where the model filters out noise by intentionally collapsing irrelevant dimensions of the feature space.

5.1.3 Automorphism

We know that the augmented vector $[1 \;\; \mathbf{x}]$ resides in the affine plane $\overline{\mathbb{X}}^p$. After the action of $\overline{\mathbf{W}}$ (from the right) on this vector, the resulting vector $[1 \;\; \mathbf{z}]$ also resides in $\overline{\mathbb{X}}^p$. Since $\overline{\mathbf{W}}$ is invertible, an inverse transformation exists, ensuring the mapping is injective in both directions:

The affine transformation is bijective.

This is known as the **automorphism** property of an affine transformation. When $p = k = 2$, an example of such a transformation is shown in Fig.2.2.

Due to this property, any transformed geometric pattern will stay within the same affine subspace. This consistency allows the network to manipulate the orientation, scale, and position of data features without destroying their underlying affine nature. We will demonstrate this behavior in detail in the examples section.

5.1.4 Preservation of Collinearity: A Line Remains a Line

An arbitrary straight line in the affine space $\mathbb{R}^{p+1}$ remains a straight line after an affine transformation. This property is fundamental to how neural networks reshape the geometry of input data.

Proof:

Consider an arbitrary line in the domain. Using a scalar parameter t, the line can be expressed as:

$$\mathbf{x}(t) = \mathbf{x}_0 + \mathbf{d}_x t \tag{5.4}$$

where $\mathbf{x}_0$ is a fixed point in $\mathbb{R}^{p+1}$ through which the line passes, and $\mathbf{d}_x$ is the constant direction (gradient) vector of the line. Applying the affine transformation from Eq.(4.3), we get:

$$\mathbf{z}(t) = \mathbf{b} + [\mathbf{x}_0 + \mathbf{d}_x t]\mathbf{W} = \underbrace{\mathbf{b} + \mathbf{x}_0\mathbf{W}}_{\mathbf{z}_0} + \underbrace{\mathbf{d}_x\mathbf{W}}_{\mathbf{d}_z} t = \mathbf{z}_0 + \mathbf{d}_z t \tag{5.5}$$

The result is a new line equation in the output space. Depending on the values within the transformation matrix, the line may be stretched, compressed, reflected, sheared, or rotated. A comprehensive list of matrices for these specific operations can be found in Ref. [22]. Additionally, the bias vector $\mathbf{b}$ ensures the line is translated.

Since the learning parameters within an affine transformation matrix are constantly updated during the training process, the affine transformation matrix will evolve to perform that specific geometric action, if the model determines that shearing or scaling the feature space is necessary to better approximate the target labels.

Singular Cases and Degeneracy: Notably, the proof for line preservation does not rely on the invertibility of $\mathbf{W}$. Therefore, this property generally holds. However, if $\mathbf{W}$ is **singular** (rank-deficient), its transpose possesses a non-trivial null-space [14].

In such cases, any line with a direction vector $\mathbf{d}_x$ that lies within the null-space of $\mathbf{W}$ (i.e., $\mathbf{d}_x\mathbf{W} = \mathbf{0}$) will collapse. Specifically, the entire line—or even higher-dimensional flat planes—will be mapped to a single point $\mathbf{z}_0$. This represents an intentional loss of dimensionality, which the network may use to filter out what it perceives as redundant information.

5.1.5 Preservation of Parallelism

Any pair of parallel lines remains parallel after an affine transformation. This property is a direct consequence of the linear nature of the transformation matrix.

Proof:

We previously established that the direction (gradient) vector of a transformed line is given by:

$$\mathbf{d}_z = \mathbf{d}_x \mathbf{W}$$

By definition, any two lines that are originally parallel share the same direction vector $\mathbf{d}_x$. When we apply the same transformation matrix $\mathbf{W}$ to both lines, their new direction vectors $\mathbf{d}_z$ will necessarily be identical. Since they share the same transformed gradient, the lines remain parallel in the output space.

If $\mathbf{W}$ is singular and the direction vector $\mathbf{d}_x$ falls within the null-space of $\mathbf{W}$, both parallel lines will collapse into two distinct points (or the same point, depending on the translation). In this case, the concept of "parallel lines" is lost to geometric degeneracy.

This preservation of parallelism is what allows neural networks to perform linear "shearing" and "stretching" while maintaining the basic structural grid of the feature space.

5.2 Invariance of Distance Ratios

5.2.1 Ratios of distances between points

The **ratios of distances** between points lying on a straight line remain unchanged under an affine transformation. This property can be formally stated as follows:

Let Eq.(4.3) be an affine transformation, and let $\mathbf{x}_1, \mathbf{x}_2, \mathbf{x}_3 \in \mathbb{R}^{(p+1)}$ be three arbitrary **distinct collinear points** such that their images under the transformation are also distinct. Then:

$$\boxed{\frac{|\mathbf{x}_1 - \mathbf{x}_2|}{|\mathbf{x}_1 - \mathbf{x}_3|} = \frac{|\mathbf{z}_1 - \mathbf{z}_2|}{|\mathbf{z}_1 - \mathbf{z}_3|}} \tag{5.6}$$

Proof:

Any arbitrary straight line can be expressed using Eq.(5.4) through a parameterization with t. For the points $\mathbf{x}_1, \mathbf{x}_2, \mathbf{x}_3$, we have:

$$\mathbf{x}_i = \mathbf{x}_0 + \mathbf{d}_x t_i, \qquad i = 1, 2, 3 \tag{5.7}$$

The left-hand side of Eq.(5.6) becomes:

$$\frac{|\mathbf{x}_1 - \mathbf{x}_2|}{|\mathbf{x}_1 - \mathbf{x}_3|} = \frac{|\mathbf{d}_x t_1 - \mathbf{d}_x t_2|}{|\mathbf{d}_x t_1 - \mathbf{d}_x t_3|} = \frac{|t_1 - t_2|}{|t_1 - t_3|} \tag{5.8}$$

Next, applying the affine transformation in Eq.(4.3), we obtain:

$$\mathbf{z}_i = \mathbf{z}_0 + \mathbf{d}_z t_i \tag{5.9}$$

where $\mathbf{z}_0 = \mathbf{b} + \mathbf{x}_0 \mathbf{W}$ and $\mathbf{d}_z = \mathbf{d}_x \mathbf{W}$. Substituting Eq.(5.9) into the right-hand side of Eq.(5.6) yields:

$$\frac{|\mathbf{z}_1 - \mathbf{z}_2|}{|\mathbf{z}_1 - \mathbf{z}_3|} = \frac{|\mathbf{d}_z t_1 - \mathbf{d}_z t_2|}{|\mathbf{d}_z t_1 - \mathbf{d}_z t_3|} = \frac{|\mathbf{d}_x \mathbf{W} t_1 - \mathbf{d}_x \mathbf{W} t_2|}{|\mathbf{d}_x \mathbf{W} t_1 - \mathbf{d}_x \mathbf{W} t_3|} \tag{5.10}$$

When $\mathbf{W}$ is non-singular, the common factor $|\mathbf{d}_x \mathbf{W}|$ cancels out, and Eq.(5.10) simplifies to:

$$\frac{|\mathbf{z}_1 - \mathbf{z}_2|}{|\mathbf{z}_1 - \mathbf{z}_3|} = \frac{|t_1 - t_2|}{|t_1 - t_3|} \tag{5.11}$$

By comparing Eq.(5.11) with Eq.(5.8), Eq.(5.6) is proven.

> **Note on Singularity:** If $\mathbf{W}$ is singular, it possesses a null space. If the direction vector $\mathbf{d}_x$ lies within this null space, the entire line collapses to a single point, rendering the ratio **undefined**.

5.2.2 ML Interpretation

In a Multi-Layer Perceptron (MLP), dimensionality remains unchanged when using an affine transformation where $p = k$ (i.e., the layer width is constant). This provides an important mechanism for **pattern detection**, as it allows the model to preserve the relative geometric structure of the input data while transforming it to match the target (label) distribution. By maintaining a constant layer width, the network can perform these affine "alignments" without losing the structural relationships defined by these distance ratios.

5.3 Mapping from Higher to Lower Dimensional Spaces

5.3.1 Dimension reduction

When $p > k$, a common occurrence in Multi-Layer Perceptrons (MLPs), the affine transformation matrix is rectangular and does not possess an inverse. In this scenario, the affine transformation maps from a higher-dimensional affine subspace to a lower-dimensional one.

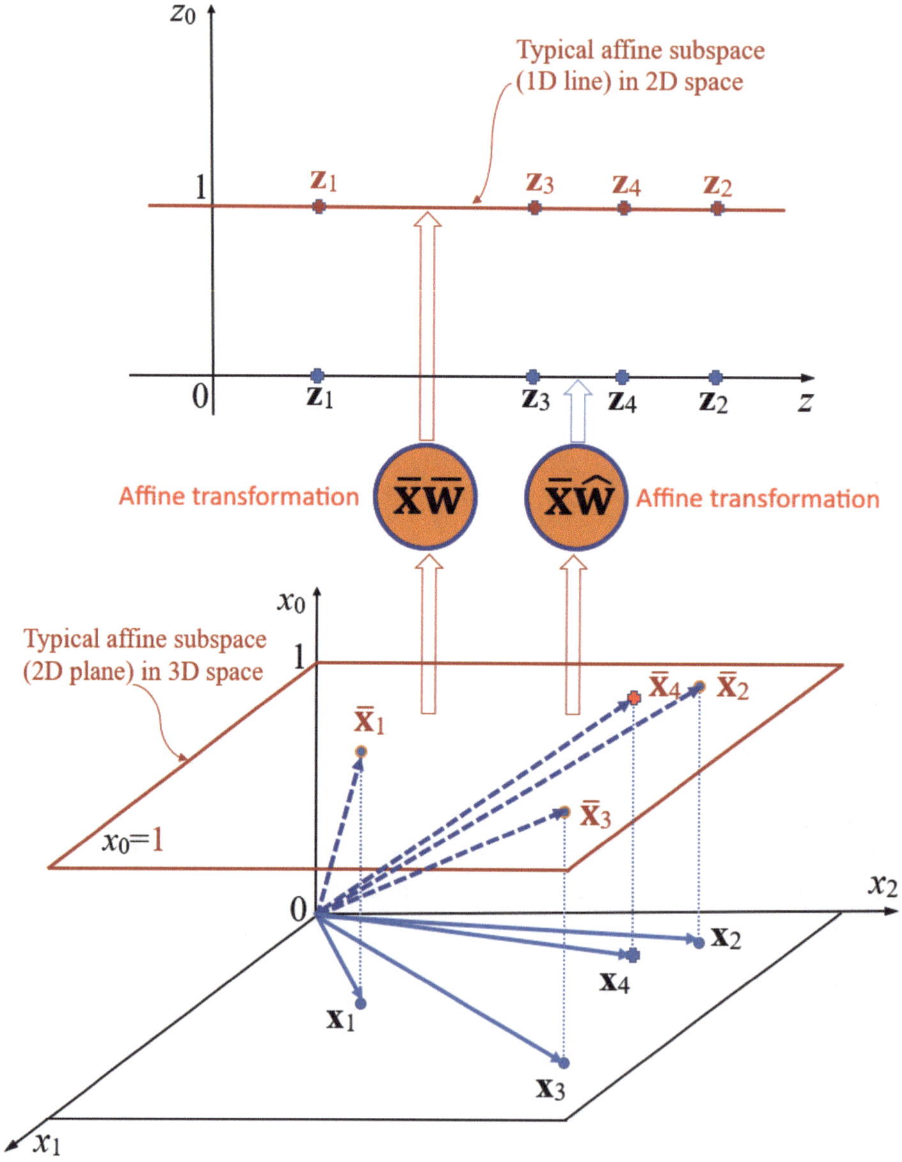

Figure 5.1: Schematic view of an affine mapping from higher affine subspace $\mathbb{X}^2$ to lower dimensional subspaces $\mathbb{X}^1$.

For example, consider the case where $p = 2$ and $k = 1$: a 2D plane in $\mathbb{R}^3$ is mapped onto a 1D line in $\mathbb{R}^2$, as shown in Fig. 5.1.

In these instances:

Dimension reduction $\Rightarrow$ non-invertible affine map

Consequently, the "affinity picture" is altered. A plane in $\overline{\mathbb{X}}^p$ may degenerate into a line or remain a plane in $\overline{\mathbb{X}}^k$, while a line in $\overline{\mathbb{X}}^p$ might collapse into a point or remain a line.

Crucially, despite the change in dimensionality, the image of an affine transformation remains an affine subspace. The core properties of **affinity**—linearity, parallelism, and distance ratios—are preserved because they are independent of the subspace's dimension.

5.3.2 ML Interpretation

In this type of affine transformation ($p > k$), dimensionality is intentionally decreased. This serves as a vital mechanism for a neural network to perform **feature selection** or **feature reduction**, which effectively facilitates **noise reduction/removal**.

By reducing the width of a layer in an MLP, the model "folds" the data manifold into a lower-dimensional representation. It is important to note that the extent of this reduction or folding is governed by the learning parameters (weights and biases), which are optimized during training to minimize the loss function.

5.4 Mapping from Lower to Higher Dimensional Spaces

When $p < k$, we encounter the **dual situation** to the dimension reduction discussed previously. In this case, the transformation maps from a lower-dimensional affine subspace to a higher-dimensional one.

For example, Fig. 5.2 illustrates $p = 1$ and $k = 2$: a 1D line in $\mathbb{R}^2$ is mapped into a 2D plane in $\mathbb{R}^3$.

5.4.1 Key Characteristics of Higher-Dimensional Mapping

While higher-dimensional targets provide more "room" for the data, the intrinsic nature of the transformation is limited by its linearity:

> Affine maps may embed or collapse—but they never expand the intrinsic dimension p because the transformation is linear.

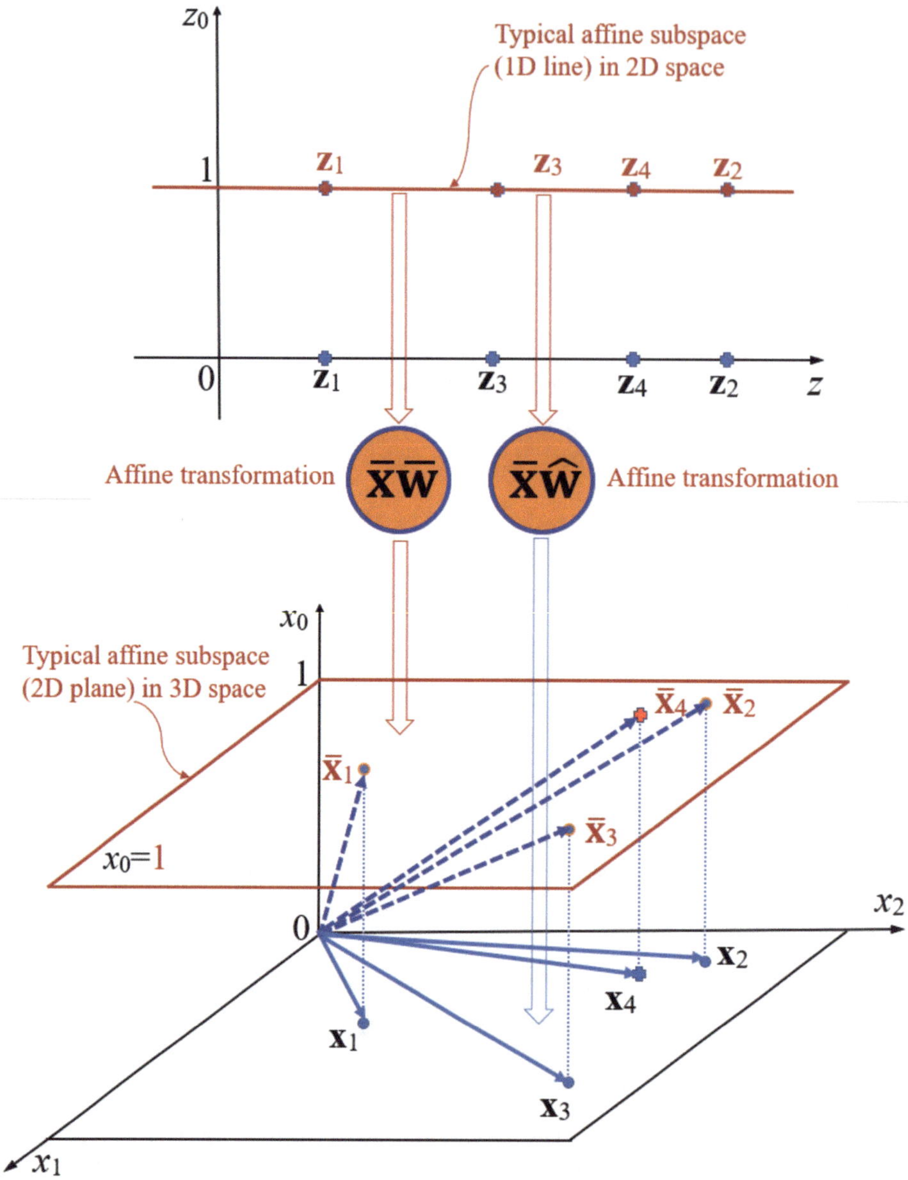

Figure 5.2: Schematic view of an affine mapping from lower affine subspace $\mathbb{X}^1$ to higher dimensional subspace $\mathbb{X}^2$.

Since the transformation matrix is not square, it does not have a standard inverse mapping. However, the core properties of **affinity** (linearity, parallelism, and distance ratios) remain preserved because they are independent of the subspace dimensions.

Specific behaviors to note:

- **Linearity Preservation:** Lines always map to lines or points—never to curves—due to the linear nature of the affine mapping.

- **Injective Mapping (Full Rank):** If $\mathbf{W}$ has full column rank, the mapping is **injective**. No two distinct points map to the same output. The domain is embedded as a p-dimensional affine subspace: $\overline{\mathbb{X}}^p \hookrightarrow \overline{\mathbb{X}}^k$. This is known as an **affine embedding**.
- **Collapsing (Rank Deficient):** If $\mathbf{W}$ is rank-deficient, it possesses a non-zero null space. In this case, the mapping is **not injective**, and vectors within the null space will collapse into a single point.

5.4.2 ML Interpretation

In an MLP, increasing the layer width ($p < k$) facilitates **feature lifting**. This is a critical mechanism for **feature expansion**, especially when followed by a non-linear activation function.

By increasing the width of a layer, the model effectively "unfolds" the data manifold into a higher-dimensional space where complex patterns might become more easily separable. As with other transformations, the specific degree of unfolding or lifting is determined by the learning parameters, which are optimized to minimize the loss function.

Chapter 6

Typical Data Structures

6.1 $p \to 1$ Networks

Equation (2.3) can be expressed in matrix form, explicitly defining the dimensionality as follows:

$$z(\hat{\mathbf{w}};\mathbf{x}) = \underset{1\times p}{\mathbf{x}}\underset{p\times 1}{\mathbf{w}} + \underset{1\times 1}{b} = \underset{1\times(p+1)}{\overline{\mathbf{x}}}\underset{(p+1)\times 1}{\hat{\mathbf{w}}}$$

The prediction function $z \in \overline{\mathbb{X}}^p$ is defined as a function of the weights $\mathbf{w}$ and bias b for any input $\mathbf{x} \in \mathbb{X}^p$. For the i-th data point $\mathbf{x}_i$, the transformation is:

$$z(\hat{\mathbf{w}};\mathbf{x}_i) = \underset{1\times p}{\mathbf{x}_i}\underset{p\times 1}{\mathbf{w}} + \underset{1\times 1}{b} = \underset{1\times(p+1)}{\overline{\mathbf{x}}_i}\underset{(p+1)\times 1}{\hat{\mathbf{w}}}$$

In this context, $z(\hat{\mathbf{w}};\mathbf{x})$ remains a scalar for an individual data point. Since this represents a single-layer network with no subsequent transformations, the augmented form $\overline{\mathbf{z}}$ is generally not required for the output.

6.1.1 Matrix Representation for Datasets

For a dataset containing m data points, the inputs are stacked vertically to form the data matrix $\mathbf{X}$. This is used to construct the matrix $\overline{\mathbf{X}}$ (the moment matrix) on the affine subspace:

$$\underset{\mathbf{z}(\hat{\mathbf{W}};\,\overline{\mathbf{X}})_{m\times 1}}{\begin{bmatrix} z_1 \\ z_2 \\ \vdots \\ z_m \end{bmatrix}} = \underset{\overline{\mathbf{X}}_{m\times(p+1)}}{\begin{bmatrix} 1 & x_{11} & \dots & x_{1p} \\ 1 & x_{21} & \dots & x_{2p} \\ \vdots & \vdots & \ddots & \vdots \\ 1 & x_{m1} & \dots & x_{mp} \end{bmatrix}} \underset{\hat{\mathbf{W}}_{(p+1)\times 1}}{\begin{bmatrix} b \\ w_1 \\ \vdots \\ w_p \end{bmatrix}} \tag{6.1}$$

Note: During actual training computations, these explicit matrices are often not formed in memory as shown. We present them here to clarify the underlying data structure and alignment.

6.1.2 Affine Subspace Transformation

If the outputs z are also required to reside on affine subspaces, the data structure is modified as follows:

$$\underbrace{\begin{bmatrix} 1 & z_1 \\ 1 & z_2 \\ \vdots & \vdots \\ 1 & z_m \end{bmatrix}}_{\overline{\mathbf{z}}(\overline{\mathbf{W}};\ \overline{\mathbf{X}})_{m\times 2}} = \underbrace{\begin{bmatrix} 1 & x_{11} & \cdots & x_{1p} \\ 1 & x_{21} & \cdots & x_{2p} \\ \vdots & \vdots & \ddots & \vdots \\ 1 & x_{m1} & \cdots & x_{mp} \end{bmatrix}}_{\overline{\mathbf{X}}_{m\times(p+1)}} \underbrace{\begin{bmatrix} 1 & b \\ 0 & w_1 \\ \vdots & \vdots \\ 0 & w_p \end{bmatrix}}_{\overline{\mathbf{W}}_{(p+1)\times 2}} \tag{6.2}$$

In this configuration, the affine transformation matrix $\overline{\mathbf{W}}$ maps (from the right) the p-dimensional affine subspace $\mathbb{X}^p \in \mathbb{R}^{(p+1)}$ to a 1D affine subspace $\mathbb{X}^1 \in \mathbb{R}^2$.

6.2 $p \to k$ Networks

In multi-dimensional cases, a layer may consist of multiple neurons, say k, in the output. Each neuron independently performs an affine transformation on the same input dataset (see Fig.4.1). Consequently, the output for a single input $\mathbf{x}_i$ is a row vector with k entries. This structure is expressed in matrix form as:

$$\underbrace{[z_1\ z_2 \cdots z_k]}_{\mathbf{z}(\mathbf{W},\mathbf{b};\ \mathbf{x}_i)_{1\times k}} = \underbrace{[x_{i1}\ x_{i2} \cdots x_{ip}]}_{\mathbf{x}_{i\ 1\times p}} \underbrace{\begin{bmatrix} w_{11} & w_{12} & \cdots & w_{1k} \\ w_{21} & w_{22} & \cdots & w_{2k} \\ \vdots & \vdots & \ddots & \vdots \\ w_{p1} & w_{p2} & \cdots & w_{pk} \end{bmatrix}}_{\mathbf{W}_{p\times k}} + \underbrace{[b_1\ b_2 \cdots b_k]}_{\mathbf{b}_{1\times k}} \tag{6.3}$$

Using augmented notation to include the bias term, the transformation becomes:

$$\underset{1\times k}{\mathbf{z}(\hat{\mathbf{W}};\ \mathbf{x}_i)} = \underset{1\times p}{\mathbf{x}_i} \underset{p\times k}{\mathbf{W}} + \underset{1\times k}{\mathbf{b}} = \underset{1\times(p+1)}{\overline{\mathbf{x}}_i} \underset{(p+1)\times k}{\hat{\mathbf{W}}} \tag{6.4}$$

Since the weights for each neuron are independent of others in the same layer, we can stack as many neurons as required. This architectural flexibility is powerful because it allows the model to approximate complex vector-valued functions.

6.2.1 $p \to k$ Networks with m Data Points

For a dataset containing m points, the data is structured into a matrix $\mathbf{X} \in \mathbb{R}^{m\times p}$ by stacking the individual input vectors $\mathbf{x}_i$ vertically. The resulting m predictions are computed simultaneously:

$$\underbrace{\begin{bmatrix} \mathbf{z}_1(\mathbf{W},\mathbf{b};\mathbf{x}_1) \\ \mathbf{z}_2(\mathbf{W},\mathbf{b};\mathbf{x}_2) \\ \vdots \\ \mathbf{z}_m(\mathbf{W},\mathbf{b};\mathbf{x}_m) \end{bmatrix}}_{\mathbf{Z}_{m\times k}(\mathbf{W},\mathbf{B};\mathbf{X})} = \underbrace{\begin{bmatrix} x_{11} & x_{12} & \dots & x_{1p} \\ x_{21} & x_{22} & \dots & x_{2p} \\ \vdots & \vdots & \ddots & \vdots \\ x_{m1} & x_{m2} & \dots & x_{mp} \end{bmatrix}}_{\mathbf{X}_{m\times p}} \underbrace{\begin{bmatrix} w_{11} & w_{12} & \dots & w_{1k} \\ w_{21} & w_{22} & \dots & w_{2k} \\ \vdots & \vdots & \ddots & \vdots \\ w_{p1} & w_{p2} & \dots & w_{pk} \end{bmatrix}}_{\mathbf{W}_{p\times k}} + \underbrace{\begin{bmatrix} \mathbf{b} \\ \mathbf{b} \\ \vdots \\ \mathbf{b} \end{bmatrix}}_{\mathbf{B}_{m\times k}} \tag{6.5}$$

Here, the matrix $\mathbf{B}$ consists of the bias vector $\mathbf{b}$ repeated for all m rows. This is concisely summarized as:

$$\underset{m\times k}{\mathbf{Z}(\hat{\mathbf{W}},\mathbf{B})} = \underset{m\times p}{\mathbf{X}}\ \underset{p\times k}{\mathbf{W}} + \underset{m\times k}{\mathbf{B}} = \underset{m\times(p+1)}{\overline{\mathbf{X}}}\ \underset{(p+1)\times k}{\hat{\mathbf{W}}} \tag{6.6}$$

6.2.2 Transformation between Affine Subspaces

If the output vectors $\mathbf{z}_i$ must also reside on an affine subspace, the structure expands to:

$$\underbrace{\begin{bmatrix} 1 & \mathbf{z}_1(\hat{\mathbf{W}};\mathbf{x}_1) \\ 1 & \mathbf{z}_2(\hat{\mathbf{W}};\mathbf{x}_2) \\ \vdots & \vdots \\ 1 & \mathbf{z}_m(\hat{\mathbf{W}};\mathbf{x}_m) \end{bmatrix}}_{\overline{\mathbf{Z}}_{m\times(k+1)}(\overline{\mathbf{W}};\overline{\mathbf{X}})} = \underbrace{\begin{bmatrix} 1 & x_{11} & \dots & x_{1p} \\ 1 & x_{21} & \dots & x_{2p} \\ \vdots & \vdots & \ddots & \vdots \\ 1 & x_{m1} & \dots & x_{mp} \end{bmatrix}}_{\overline{\mathbf{X}}_{m\times(p+1)}} \underbrace{\begin{bmatrix} 1 & b_1 & \dots & b_k \\ 0 & w_{11} & \dots & w_{1k} \\ \vdots & \vdots & \ddots & \vdots \\ 0 & w_{p1} & \dots & w_{pk} \end{bmatrix}}_{\overline{\mathbf{W}}_{(p+1)\times(k+1)}} \tag{6.7}$$

Or more simply:

$$\underset{m\times(k+1)}{\overline{\mathbf{Z}}} = \underset{m\times(p+1)}{\overline{\mathbf{X}}}\ \underset{(p+1)\times(k+1)}{\overline{\mathbf{W}}} \tag{6.8}$$

In this scenario, the transformation matrix $\overline{\mathbf{W}}$ maps a p-dimensional affine subspace $\mathbb{X}^p \subset \mathbb{R}^{(p+1)}$ to a k-dimensional affine subspace $\mathbb{X}^k \subset \mathbb{R}^{(k+1)}$.

Implementation Note: In practice, these large matrices are rarely explicitly instantiated as shown. During training, loss functions are typically computed as summations over individual samples or mini-batches. These representations serve primarily to visualize the data flow within the model.

Chapter 7

Demonstration of Affine Transformations in 1D

To examine the geometric effects of affine transformations, we construct a general transformation matrix $\mathbf{A} \in \mathbb{R}^{2\times 2}$. The matrix $\mathbf{A}$ is deliberately chosen to be non-singular, which guarantees that the resulting mappings are injective (one-to-one).

In the one-dimensional affine setting, input vectors are represented in the augmented space $\mathbb{R}^2$ in order to incorporate the translation (bias) term. This augmented representation allows both linear transformation and translation to be expressed compactly as a single matrix operation.

To visualize how vectors in the augmented space are transformed by the matrix $\mathbf{A}$, we employ a set of Python utility functions. These functions are used to draw and compare the original vectors and their transformed counterparts, thereby illustrating the geometric action of the affine transformation.

```
from matplotlib.ticker import MultipleLocator
plt.rcParams['figure.dpi'] = 300

def drawArrow(axis, origin, vector, color='k', iax=0, xylim=6,\
                                                  label=None):
    aw, lw = 0.015*xylim, 0.09*xylim
    # handles both labeled and unlabeled cases
    arr = axis[iax].arrow(*origin, *vector, width=aw, lw=lw,
                          color=color, fill=False, label=label,
                          length_includes_head=True)
    return arr

def plotfig(axis, iax=0, title="f_title", xylim=6, spacing=1):
    ax = axis[iax] # Create a local reference for brevity
```

```
    ax.set_axisbelow(True) # Grid and Ticks
    locator = MultipleLocator(spacing)
    ax.xaxis.set_minor_locator(locator)
    ax.yaxis.set_minor_locator(locator)
    ax.grid(ls=':', lw=0.5, which='both')

    ax.axis('square') # Plotting and Styling
    ax.plot(0, 0, 'ro', markersize=2)
    ax.set(xlim=(-xylim, xylim), ylim=(-xylim, xylim), title=title)
    ax.legend()
```

The following snippet illustrates vectors in two-dimensional (2D) space after being transformed by left multiplication with an affine matrix $\mathbf{A} = \overline{\mathbf{W}}^\top$. For comparison, we include both the standard basis vectors (SBVs, $\mathbf{i}_1$ and $\mathbf{i}_2$) and two arbitrary vectors,$\mathbf{v}_1$ and $\mathbf{v}_2$.

```
# 1. Setup Data
o1 = np.array([0, 0]) # origin
# Basis and arbitrary vectors organized into lists for easy iteration
vecs_left = [np.array([1, 0]), np.array([0, 1])] # base vectors i1,i2
vecs_right= [np.array([1,.5]), np.array([1,-.8])]  # arbitrary v1, v2

A = np.array([[1, 0], [0.7, 1.1]]) # affine matrix (acts from left)
print(f'An affine matrix A for acting from the left:\n{A}')
print(f'The rank of A: {np.linalg.matrix_rank(A)}')

# 2. Plotting Configuration
fig, axis = plt.subplots(1, 2, figsize=(8, 4))
titles = ['Transformed SBVs', 'Transformed Arbitrary Vectors']
labels_base = [['i1', 'i2'], ['v1', 'v2']]

colors = ['r', 'm', 'b', 'c']
# Original, Original, Transformed, Transformed

# 3. Unified Plotting Loop
for i, (vectors, title) in enumerate(\
                              zip([vecs_left, vecs_right], titles)):

    xylm, mg = 1.5, (2 if i==0 else 1.5)
    handles = []

    for j, v in enumerate(vectors):
        Av = A.dot(v)
```

```
        # Draw Original Vector
        handles.append(drawArrow(axis, o1, v, color=colors[j], iax=i,
                                  xylim=mg*xylm, label=labels_base[i][j]))
        # Draw Transformed Vector
        handles.append(drawArrow(axis, o1,␣
↪Av,color=colors[j+2],iax=i,
                         xylim=mg*xylm, label=f'A@{labels_base[i][j]}'))

        print(f"{labels_base[i][j]}={v}, A@{labels_base[i][j]}={Av}")

    axis[i].legend(handles=handles, loc='upper left')
    plotfig(axis, iax=i, title=title, xylim=xylm)

plt.tight_layout()
plt.show()
```

```
An affine matrix A for acting from the left:
[[1.  0. ]
 [0.7 1.1]]
The rank of A: 2
i1=[1 0], A@i1=[1.  0.7]
i2=[0 1], A@i2=[0.  1.1]
v1=[1.  0.5], A@v1=[1.   1.25]
v2=[ 1.  -0.8], A@v2=[ 1.   -0.18]
```

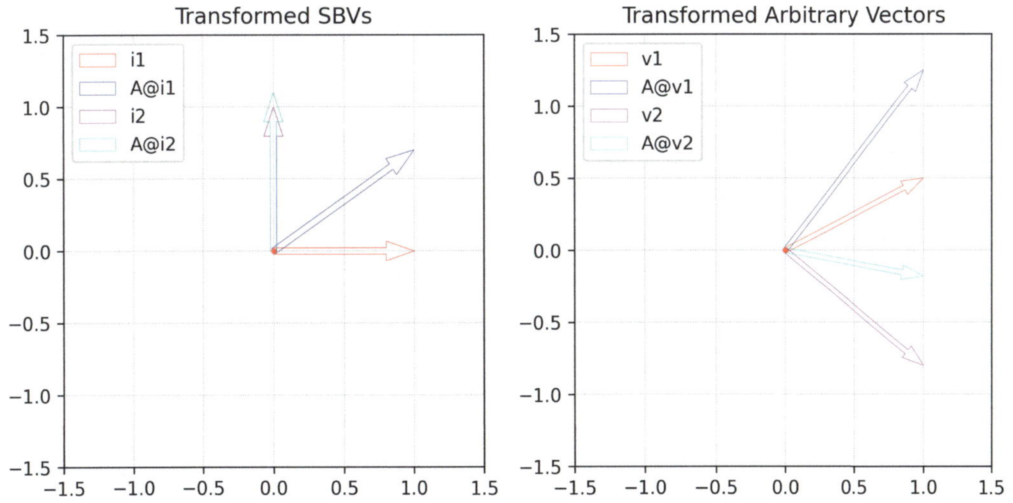

It is observed that the first component of all vectors (except the SBV $\mathbf{i}_2$, whose first component is zero) is mapped to 1 by the affine transformation matrix. As a result, these vectors lie on the same affine subspace. This illustrates the automorphism property of affine transformations.

Chapter 8

Demonstration of Affine Transformation in 2D

In this section, we present several examples of affine transformations applied to geometric patterns. The transformation procedure is described as follows.

Consider a geometric pattern represented by a set of data points

$$\overline{\mathbf{X}} \in \overline{\mathbb{X}}^2 \tag{8.1}$$

which lies in a two-dimensional affine subspace. Each data point corresponds to a row of $\overline{\mathbf{X}}$ and is written in augmented form as

$$\overline{\mathbf{x}}_i = [1, x_{i1}, x_{i2}] \tag{8.2}$$

The affine transformation is computed using Eq. (2.4) to produce two separate outputs.

First, we define

$$z_I = \overline{\mathbf{X}}\,\hat{\mathbf{w}}_I \tag{8.3}$$

where $\hat{\mathbf{w}}_I = [b_I, \mathbf{w}_I]^\top$, and b_I and $\mathbf{w}_I$ denote a specific set of learning parameters in the hypothesis space $\mathbb{R}^3$.

Similarly, a second output is obtained as

$$z_{II} = \overline{\mathbf{X}}\,\hat{\mathbf{w}}_{II} \tag{8.4}$$

where $\hat{\mathbf{w}}_{II} = [b_{II}, \mathbf{w}_{II}]^\top$ represents a different choice of learning parameters in $\mathbb{R}^3$.

Together, these two outputs form the transformed dataset

$$\overline{\mathbf{Z}} = [1, z_I, z_{II}] \in \overline{\mathbb{X}}^2 \tag{8.5}$$

By applying the affine transformation to the original dataset $\overline{\mathbf{X}}$ twice—using distinct parameter vectors $\hat{\mathbf{w}}_I$ and $\hat{\mathbf{w}}_{II}$—we generate a new dataset $\overline{\mathbf{Z}}$ that remains within the same affine subspace $\overline{\mathbb{X}}^2$. Such a mapping, which preserves the structure of the affine subspace (automorphism).

To analyze the **affine nature** of this transformation, we visualize the results in a two-dimensional plane. Since $\overline{\mathbb{X}}^2$ is itself a 2D plane, both the original and transformed patterns can be plotted using only the coordinates (z_I, z_{II}). This representation provides a clear geometric interpretation of the transformation without requiring a three-dimensional visualization.

8.1 Utility Functions

To demonstrate these concepts computationally, we first introduce the necessary tools. The following code snippets define the activation functions and network structures used in the subsequent analysis.

```
def logistic(z): # The sigmoid/logistic function
    return 1. / (1. + np.exp(-z))

def relu(z):       # The ReLU activation function
    return np.maximum(0.0, z)

def tanh(z):       # The tanh activation function
    return np.tanh(z)     # use the numpy.tanh()

def net(x_bar,w_hat): # net without activation (linear activation)
    '''An affine transformation net.
    x should be given in the form of x_bar (with 1 at the 1st entry)
    w should be given in the form of w_hat (with b on top)'''
    z = np.dot(x_bar, w_hat)
    return z

def net_σ(x_bar,w_hat):
    '''An affine transformation net with logistic activation'''
    z = np.dot(x_bar, w_hat)
    return logistic(z)

def net_relu(x_bar,w_hat):
```

```
    '''An affine transformation net with relu activation'''
    z = np.dot(x_bar, w_hat)
    return relu(z)

def net_tanh(x_bar,w_hat):
    '''An affine transformation net with relu activation'''
    z = np.dot(x_bar, w_hat)
    return tanh(z)
```

The following snippets define some objects of different geometric patterns.

```
def edge(k, dd):
    """Generates a 2D line (x2 = k*x1) with an augmented
       intercept column.
    """
    # Create the base range including the endpoint
    x1 = np.arange(-1.0, 1.0 + dd, dd)

    # Use array concatenation/stacking directly for efficiency
    x1 = np.append(x1, 0) # Ensure center (0,0) is included
    x2 = k * x1

    # Construct the design matrix X: [1, x1, x2]
    X = np.column_stack([np.ones_like(x1), x1, x2])

    return x1, x2, X

def circle(r, dd):
    """Generates a circular pattern in 2D feature space with an
       augmented column.
    """
    # Generate angles and coordinates
    theta = np.arange(0.0, 2*np.pi, dd)
    x1, x2 = r * np.cos(theta), r * np.sin(theta)

    # Append the center (0,0) to both coordinates
    x1, x2 = np.append(x1, 0), np.append(x2, 0)

    # Create the augmented design matrix X: [1, x1, x2]
    X_bar = np.column_stack([np.ones_like(x1), x1, x2])

    return x1, x2, X_bar

def rectangle(MinX, MaxX, MinY, MaxY, dd, theta):
```

```
    """Generates a rotated rectangular boundary with an augmented
       center point. """

    # Define the four edges of the rectangle
    x_range = np.arange(MinX, MaxX + dd, dd)
    y_range = np.arange(MinY, MaxY + dd, dd)

    # Coords. for Bottom, Right, Top (flipped),& Left (flipped) edges
    x1 = np.concatenate([x_range, np.full_like(y_range, MaxX), \
                         x_range[::-1], np.full_like(y_range, MinX)])
    x2 = np.concatenate([np.full_like(x_range, MinY), y_range, \
                         np.full_like(x_range, MaxY), y_range[::-1]])

    # Add the center point
    x1 = np.append(x1, (MinX + MaxX) / 2)
    x2 = np.append(x2, (MinY + MaxY) / 2)

    # Apply rotation transformation
    X1 = x1 * np.cos(theta) + x2 * np.sin(theta)
    X2 = x2 * np.cos(theta) - x1 * np.sin(theta)

    # Create augmented design matrix [1, X1, X2]
    X = np.column_stack([np.ones_like(X1), X1, X2])

    return X1, X2, X

def spiral(alpha, c, xright=40.01, dd=0.05):
    """Generates a logarithmic spiral pattern with an augmented
       column.
    """
    # Define range; include xright in the array
    t = np.arange(0, xright, dd)

    # Calculate spiral coordinates: r = exp(alpha * t) / c
    decay = np.exp(alpha * t) / c
    x1, x2 = decay * np.cos(t), decay * np.sin(t)

    # Append the center (0,0) and create augmented matrix [1, x1, x2]
    x1, x2 = np.append(x1, 0), np.append(x2, 0)
    X = np.column_stack([np.ones_like(x1), x1, x2])

    return x1, x2, X
```

8.2 Manual Setting Learning Parameters

Let us now set up learning parameters $\mathbf{w}$ and b.

```
def to_affine(w, b=0):
    """Helper to stack bias and weights into an augmented vector."""
    return np.vstack(([b], w))

# 1. Initialize SBVs (Standard Basis Vectors) as column vectors
w0_I  = np.array([[1.], [0.]])
w0_II = np.array([[0.], [1.]])

# 2. Augmented vectors (Identity transformation: z_I = x1, z_II = x2)
wh0_I  = to_affine(w0_I,  b=0)      # w_hat
wh0_II = to_affine(w0_II, b=0)      # w_hat

Wh0 = np.hstack((wh0_I, wh0_II))    # with identity matrix as: W_hat_0
# Display results
#print(f"w_hat, wh0_I:\n{wh0_I}\nw_hat, wh0_II:\n{wh0_II}")
print(f"Speical affine with identity matrix: \n{Wh0}") # shape:(3, 2)
```

```
Speical affine with identity matrix:
[[0. 0.]
 [1. 0.]
 [0. 1.]]
```

```
# 1. Define Weights and Biases
# Using shorthand for column vectors: .reshape(-1, 1) or [[val]]
w_I  = np.array([[0.8], [0.2]])
w_II = np.array([[0.2], [0.5]])
b_I  = np.array([[-.6]])
b_II = np.array([[0.2]])
b0   = np.array([[0.0]])  # Standard zero bias

# 2. Build Augmented Weight Vectors (w_hat)
# vstacking bias (intercept) with weights
wh_I      = np.vstack((b_I,  w_I))
wh_II     = np.vstack((b_II, w_II))
wh_I_b0   = np.vstack((b0,   w_I))
wh_II_b0  = np.vstack((b0,   w_II))

W_b0 = np.hstack((wh_I_b0, wh_II_b0)) # linear transformation matrix
W_hat = np.hstack((wh_I_b0, wh_II))   # Affine transformation matrix
```

```
#print(f"Affine wh_I_b0:  \n{wh_I_b0} ") # shape: (3, 1)
#print(f"Affine wh_II:    \n{wh_II}")    # shape: (3, 1)
print(f"Linear transformation matrix W:  \n{W_b0}")  # shape: (3, 2)
print(f"Affine transformation matrix Wh: \n{W_hat}") # shape: (3, 2)
```

```
Linear transformation matrix W:
[[0.  0. ]
 [0.8 0.2]
 [0.2 0.5]]
Affine transformation matrix Wh:
[[0.  0.2]
 [0.8 0.2]
 [0.2 0.5]]
```

8.3 Tool to Plot Affine-Transformed Patterns

Next, we define a function to plot the patterns in three forms: the original patterns, their affine-transformed versions using $\mathbf{w}_{II}$ and b_{II}, and their linearly transformed counterparts using $\mathbf{w}_{II}$ with $b_{II} = 0$.

```
def affineplot(net, X, wh0_I, wh0_II, wh_I, wh_II, wh_I_b0,␣
 ↪wh_II_b0):
    """Plots and compares original, affine, and linear
       transformations.
    """
    plt.figure(figsize=(8, 8), dpi=200)
    # Helper to generate legend labels from weight arrays
    def get_label(prefix, w, b):
        return f"{prefix}: w=[{w[1]},{w[2]}], b={b}"

    # 1. Original Pattern
    plt.scatter(net(X, wh0_I), net(X, wh0_II), s=10, c='orange',
                label=f"{get_label('Original', wh0_I, wh0_I[0])}\n"
                      f"{get_label('        ', wh0_II, wh0_II[0])}")

    # 2. Affine Transformed Pattern
    plt.scatter(net(X, wh_I), net(X, wh_II), s=10, c='blue',
                label=f"{get_label('Affine  ', wh_I, wh_I_b0[0])}\n"
                      f"{get_label('        ', wh_II, wh_II[0])}")

    # 3. Linear Transformed Pattern
    plt.scatter(net(X, wh_I_b0), net(X, wh_II_b0), s=10, c='red',
                label=f"{get_label('Linear  ', wh_I, wh_I_b0[0])}\n"
                      f"{get_label('        ', wh_II, wh0_II[0])}")
```

```
# Formatting
plt.xlabel('$z_{I}$')
plt.ylabel('$z_{II}$')
#plt.title('Linear and Affine Transformation')
plt.grid(color='r', ls=':', lw=0.3)
#plt.legend(loc='center left', bbox_to_anchor=(1, 0.5),\
#                            prop={'family': 'monospace'})
plt.axis('scaled')

plt.savefig('images_XW/learn_affine.png', dpi=500,\
                                     bbox_inches='tight')
plt.show()
```

8.4 Plots of Affine-Transformed Patterns

8.4.1 An Edge under Affine Transformation

We first construct an edge (a straight line segment) represented by a set of orange points. We then apply the affine transformation defined in Eq. (2.4) and plot the patterns before transformation, as well as after the affine (blue) and linear (red) transformations.

```
x1, x2, Xb = edge(1.0, 0.1)
affineplot(net, Xb, wh0_I, wh0_II, wh_I, wh_II, wh_I_b0, wh_II_b0)
```

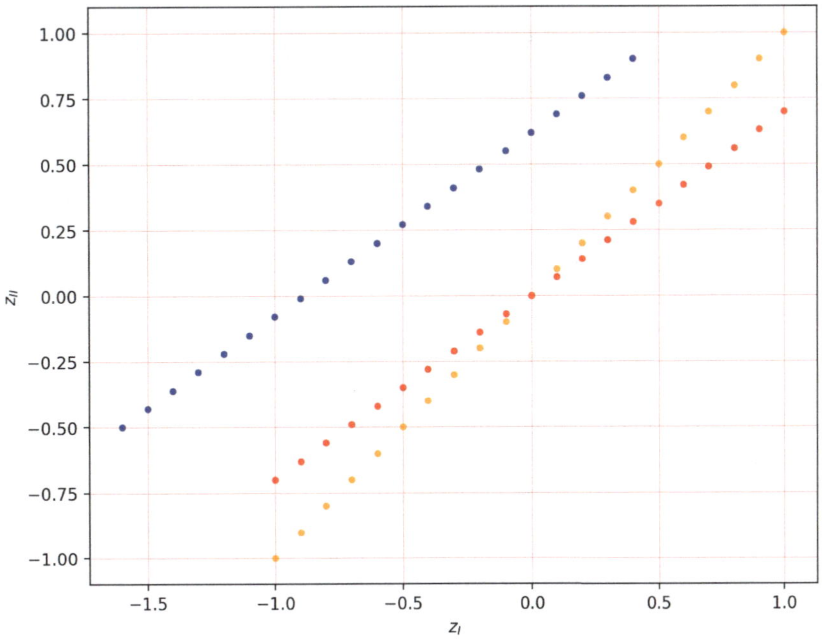

From the above plot, the following observations can be made:

1. The line transformed by the linear mapping (red) is a rotated version of the original line (orange).
2. The line transformed by the affine mapping (blue) is a rotated and translated version of the original line (orange).
3. Distances between points are not preserved by the transformations; only the ratios of distances between points lying on the same straight line are preserved.

8.4.2 A Rectangle under Affine Transformation

We now construct a rectangular pattern represented by a set of orange points. We then apply the transformations and plot the pattern before transformation, as well as after the affine (blue) and linear (red) transformations. Since a rectangle is composed of straight line segments, the preservation of affinity under the transformation is easy to observe.

```
x1, x2, Xb = rectangle(-2.,2.,-1.,1.,0.1,np.pi/4)
affineplot(net, Xb, wh0_I, wh0_II, wh_I, wh_II, wh_I_b0, wh_II_b0)
```

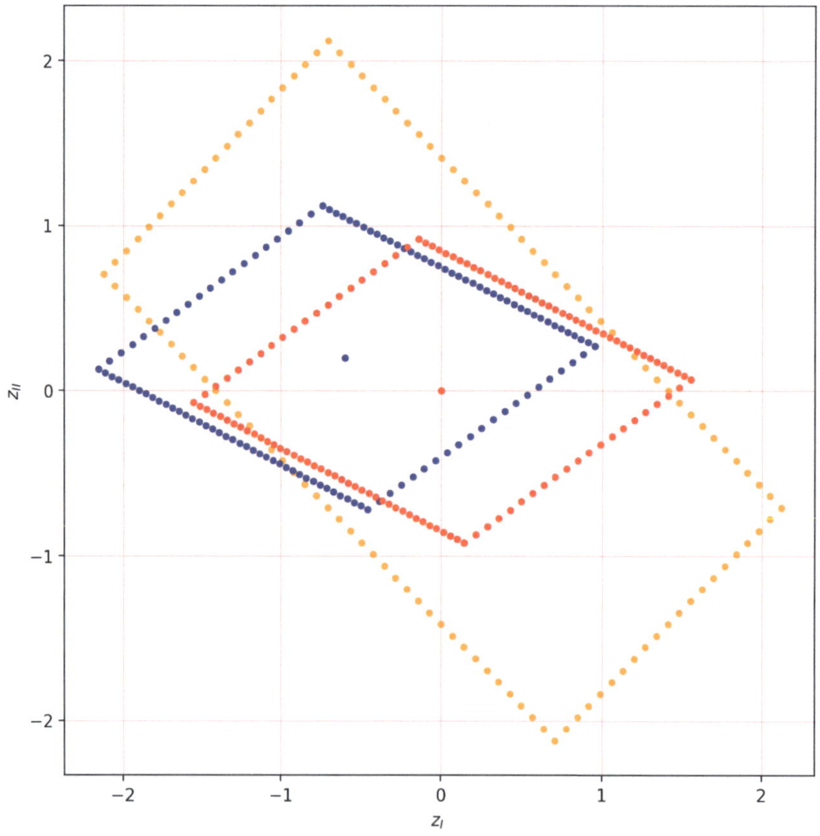

From the above plot, the following observations can be made:

1. After the affine transformation, the original (orange) rectangular pattern is rotated, scaled, sheared, and translated into a new one (blue). The matrix **w** is responsible for the linear transformation, which produces scaling, rotation, and shearing, while b is responsible for translation.

2. The transformation maps points to points, edges to edges, and quadrilaterals to quadrilaterals.

3. The transformation preserves the ratios of the lengths of parallel line segments. For example, the ratio of the two longer sides of the orange rectangle is the same as the ratio of the two longer sides of the blue quadrilateral.

4. Parallel line segments remain parallel after the affine transformation.

5. Distances between points are not preserved. Only the ratios of distances between points lying on the same straight line are preserved.

6. Angles between lines are not preserved under the affine transformation.

This simple demonstration helps illustrate the affinity of affine transformations. Readers can visualize how a pattern is mapped, and how the mapped patterns can cover the (same) space by varying **w** and b. A pure linear transformation alone does not shift the origin and therefore has much more limited coverage.

8.4.3 A Circle under Affine Transformation

Let us now examine the affine and linear transformations of a circle using the same code.

```
x1, x2, Xb = circle(1.0,0.05)
affineplot(net, Xb, wh0_I, wh0_II, wh_I, wh_II, wh_I_b0, wh_II_b0)
```

This time, it is clearly seen from the above figure that, after the transformation, the original (orange) circular pattern is rotated, scaled, sheared, and translated into an ellipse (blue). The observations made for the rectangular patterns remain valid in this case.

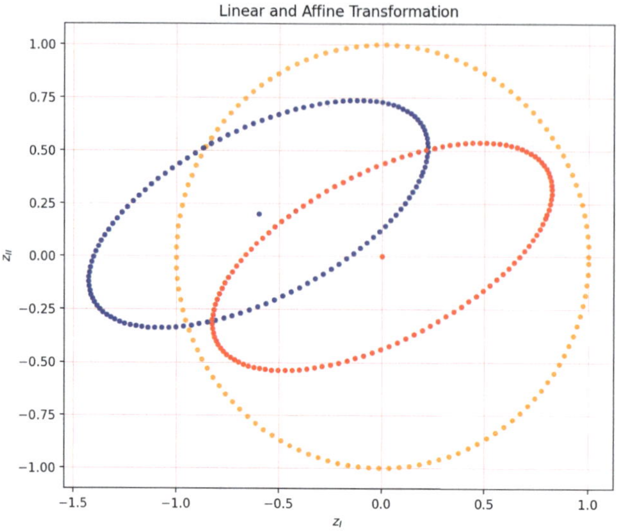

8.4.4 A Spiral under Affine Transformation

Let us now examine the affine and linear transformations of a more complex pattern—a spiral.

```
x1, x2, Xb = spiral(0.1,10.)
affineplot(net, Xb, wh0_I, wh0_II, wh_I, wh_II, wh_I_b0, wh_II_b0)
```

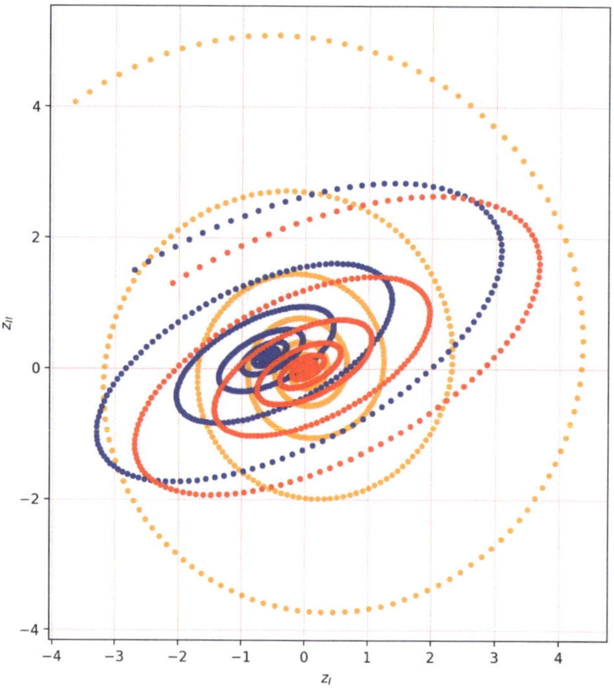

In this case, the original (orange) spiral is rotated, scaled, sheared, and translated to produce a new pattern. The observations made in the previous examples remain valid.

8.4.5 Fern Leaf under Affine Transformation

Figure 8.1 provides a compelling illustration of affine transformations as they appear in nature [22]. Individual leaf structures within a fern are related through affine transformations that combine rotation, scaling, reflection, and translation. In particular, the red leaf branch represents an affine transformation of the entire fern leaf, as well as of the dark blue and light blue leaf branches, and even of sub-branches within those structures.

Figure 8.1: A leaf of the fern-like fractal is an affine transformation of another, exhibiting self-similarity both within the same hierarchical level and across different hierarchical levels. The image at the top shows a photograph of an actual fern leaf, while the image at the bottom depicts a fractal representation of Barnsley's fern. The latter is adapted from a public-domain source by António Miguel de Campos.

This relationship demonstrates the inherent self-similarity of the fractal, both within the same hierarchical level and across different hierarchical levels. Such self-similarity can be formally represented and systematically generated using affine transformations.

The repeating geometric patterns observed in the fern exemplify the fundamental role of affine transformations in producing complex, self-similar structures from simple transformation rules.

Chapter 9

Activated Affine Transformation

9.1 On the Role of Nonlinear Activation

An affine transformation is inherently linear and, when used in isolation, is limited to modeling simple linear relationships. To address the nonlinear nature of real-world problems, affine transformations are wrapped with a nonlinear function known as an **activation function**, denoted by $\phi(\cdot)$. This architectural concept is illustrated in Fig.3.1, Fig.4.1, and Fig.4.2.

Although a wide variety of activation functions have been proposed in the machine learning literature, the most commonly used include **ReLU** (Rectified Linear Unit), **tanh** (hyperbolic tangent), and **logistic** (or sigmoid). The following simple snippet plots these three functions.

```
z = np.linspace(-5, 5, 500)          # Define input range for the␣
 ↪plot

reluz = relu(z)
reluz[z > 2] = np.nan  # remove values beyond x=3; no horizontal line

plt.figure(figsize=(8,5))                   # Plot the activation functions
plt.plot(z, reluz, label='ReLU $[0, \infty)$',c='red', linewidth=2)
plt.plot(z, tanh(z), label='tanh $(-1, 1)$', c='blue', linewidth=2)
plt.plot(z, logistic(z), label='logistic $(0, 1)$',c='green',\
                                                        linewidth=2)
plt.title('Activation Functions') # Labels and grid
plt.xlabel('$z$')
plt.ylabel(r'$\phi(z)$')
plt.grid(True)
plt.legend(loc="upper left")
plt.ylim([-1.2, 2.0])
```

```
plt.show()
```

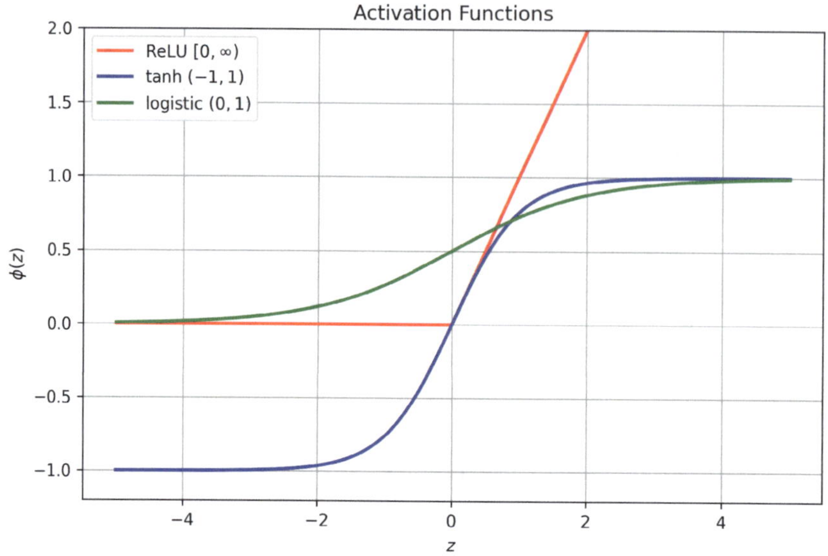

The characteristics of these activation functions are summarized in the following table.

Activation	Formula	Output Range	Key Features / Advantages	Typical Use
ReLU	$\max(0, x)$	$[0, \infty)$	Simple, avoids vanishing gradients for positive inputs, computationally efficient	Hidden layers in feedforward and convolutional networks
tanh	$\frac{e^x - e^{-x}}{e^x + e^{-x}}$	$(-1, 1)$	Zero-centered, smooth, differentiable, helps optimization	Recurrent neural networks, hidden layers requiring negative activations
Sigmoid	$\frac{1}{1+e^{-x}}$	$(0, 1)$	Smooth, bounded, interpretable as probability; but can saturate and cause vanishing gradients	Output layer for binary classification

9.2 Examples: Affine Transformation Wrapped with Activation Function

When an affine transformation $z(\mathbf{w}, b; \mathbf{x})$ is wrapped with a nonlinear activation function, the resulting output $\phi(z)$ is governed by the characteristics of that activation.

As a consequence, the **affinity**—which we previously showed preserves linearity and distance ratios—is generally destroyed at the global level.

Nevertheless, because activation functions used in machine learning are continuous and smooth (or at least piecewise differentiable), the composed mapping $\phi(z)$ acquires the essential ability to approximate nonlinear functions.

In this section, we examine the geometric patterns produced when subjected to an affine transformation followed by an activation function. For comparison, we focus on two representative activation functions: **ReLU** and **tanh**.

9.2.1 An Edge under Activated Affine Transformation

The following code snippet illustrates the transformation of a straight line when wrapped with the **ReLU** activation function. The ReLU maps a domain of $(-\infty, \infty)$ to a co-domain of $[0, \infty)$, producing a characteristic piecewise-linear deformation of the original edge.

```
x1, x2, Xb = edge(1.0, 0.02)
print(f"Linear transformation matrix: \n{W_b0}") #[wh_I_b0, wh_II_b0]
print(f"Affine transformation matrix: \n{W_hat}")   # [wh_I_b0 wh_II]
affineplot(net_relu, Xb, wh0_I, wh0_II, wh_I, wh_II,wh_I_b0,wh_II_b0)
```

```
Linear transformation matrix:
[[0.  0. ]
 [0.8 0.2]
 [0.2 0.5]]
Affine transformation matrix:
[[0.  0.2]
 [0.8 0.2]
 [0.2 0.5]]
```

The plot is shown in the following page.

Comparing these results with those of pure affine transformations, we observe that the linear transformation (red) simply rotates the original line (orange). In this particular case, the outcome is identical to that of an affine transformation without an activation function. This occurs because, over the plotted domain, the transformed line remains entirely positive; consequently, the ReLU activation—which only modifies negative values—has no effect on the output.

As long as the affine mapping projects the input data into the first quadrant (or, more generally, into the positive range of the activation function), global affinity is preserved. The characteristic "folding" or "hinging" behavior of the model manifold emerges only when the transformation shifts or rotates part of the data into the negative domain of the hidden neurons.

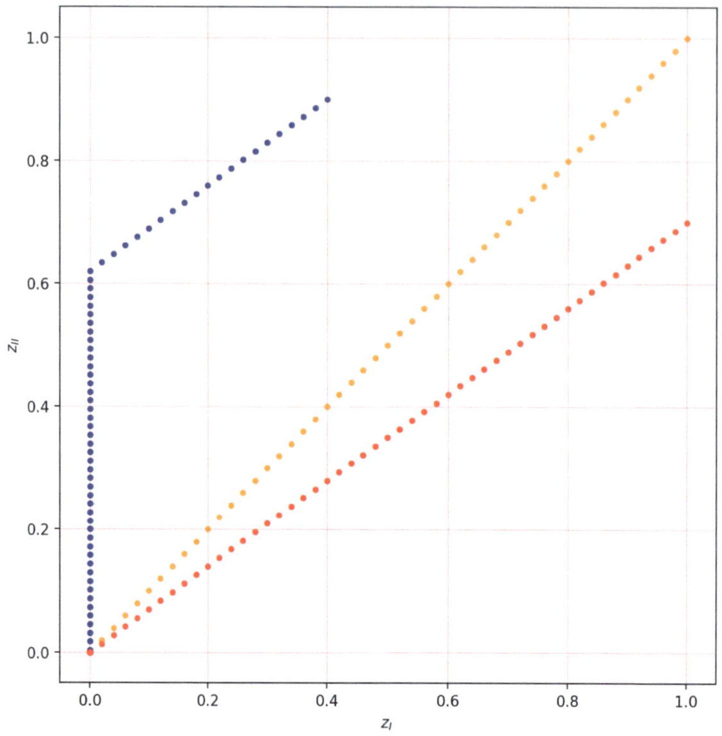

This behavior is observed for the affine-transformed line (blue), where the z_I values contain a negative segment within the domain. After applying the ReLU activation, this negative portion is mapped to zero, appearing as a horizontal segment. This clearly illustrates the **piecewise linear** nature of ReLU: one portion of the line is transformed as if connected by a hinge. At the hinge point, the line remains continuous, but its slope changes abruptly. As a result, a model manifold constructed using ReLU can exhibit points at which gradients are discontinuous, even though the manifold itself remains continuous.

The following snippet plots the affine-transformed straight line wrapped with the **tanh** activation function, which maps a domain of $(-\infty, \infty)$ to a co-domain of $(-1, 1)$.

```
x1, x2, Xb = edge(1.0,0.02)
affineplot(net_tanh, Xb, wh0_I, wh0_II, wh_I, wh_II,wh_I_b0,wh_II_b0)
```

The plot is shown in the following page.

Comparing this with the results of pure affine transformations—under which all geometric shapes retain their straight-edged characteristics—the **tanh-wrapped** affine transformation warps these lines into smooth curves. Because the tanh function is infinitely differentiable, the resulting model manifold is expected to be continuously curved and smooth throughout the space.

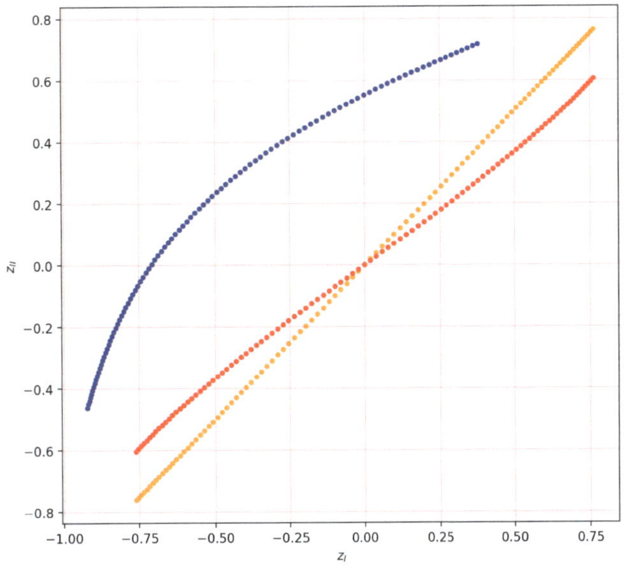

9.2.2 A Rectangle under Activated Affine Transformation

The following snippet plots an affine-transformed rectangle wrapped with the **ReLU** activation function:

```
x1, x2, Xb = rectangle(-2.,2.,-1.,1.,0.05,np.pi/4)
affineplot(net_relu, Xb, wh0_I, wh0_II, wh_I, wh_II,wh_I_b0,wh_II_b0)
```

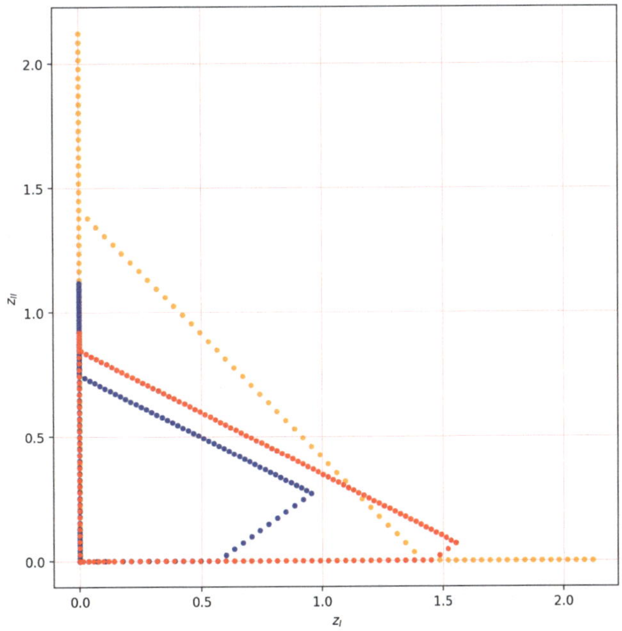

As shown, all geometric structures—the original, linearly transformed, and affine-transformed rectangles—are mapped into connected line segments under the action of ReLU. The characteristic "folding" effect arises where the transformation pushes portions of the rectangle across the coordinate axes, causing those segments to collapse onto the boundaries of the positive quadrant.

For comparison, the following snippet plots the affine-transformed rectangle wrapped with the **tanh** activation function:

```
x1, x2, Xb = rectangle(-2.,2.,-1.,1.,0.05,np.pi/4)
affineplot(net_tanh, Xb, wh0_I, wh0_II, wh_I, wh_II,wh_I_b0,wh_II_b0)
```

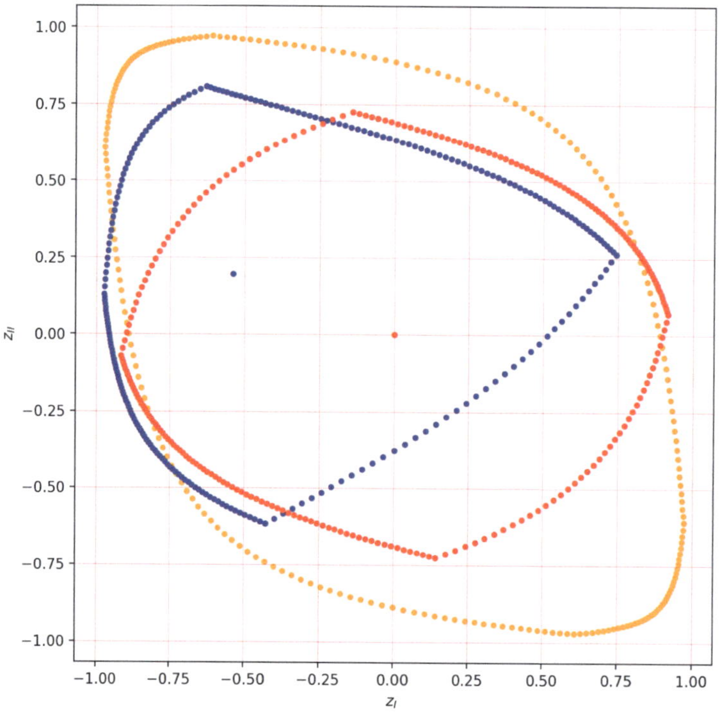

In this case, all the rectangles are transformed into smooth, continuous loops. The loops maintain continuity throughout, except at the original corners of the rectangles, where the mapping preserves a piecewise character.

9.2.3 A Circle under Activated Affine Transformation

Unlike rectangles, which are defined by discrete vertices and straight edges, a circle provides a continuous set of points—making it ideal for testing the behavior of the manifold. Under a pure affine transformation, a circle is mapped to an ellipse, preserving its "conic" affinity, as shown previously. However, when an activation function is applied, the geometric properties of the mapping change fundamentally.

The following snippet plots the affine-transformed circle wrapped with the **ReLU** activation function:

```
x1, x2, Xb = circle(1.0, 0.02)
affineplot(net_relu, Xb, wh0_I, wh0_II, wh_I, wh_II,wh_I_b0,wh_II_b0)
```

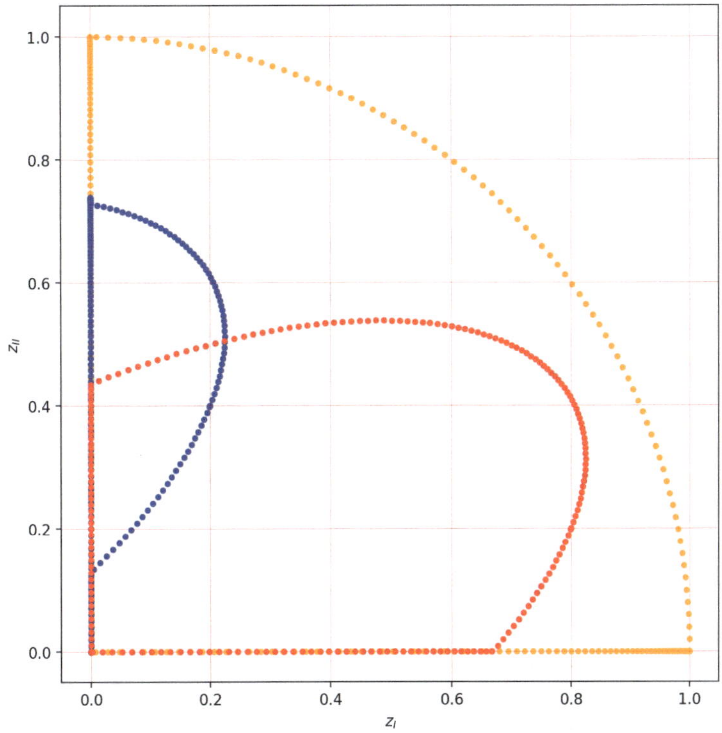

Compared to the results of pure affine transformations—where the original circle is simply mapped to translated ellipses—the **ReLU-wrapped** affine transformation preserves only the portions of the ellipse that lie in the positive domain. The remaining segments are clipped, collapsing onto the z-axes and producing a shape with flattened edges.

The following Python function can be used to plot the original pattern, the affine-transformed pattern, and the transformed pattern wrapped with an activation function.

```
def plot_transformation(x1, x2, X_bar, W_hat, activation_fn=relu,\
                                          title="Manifold Transformation"):
    """
    Works directly with the output of circle(r, dd).
    """
    # 1. Compute Transformations using the augmented X_bar
    Z_linear = net(X_bar, W_hat)
    Z_activated = activation_fn(Z_linear)

    # 2. Visualization
```

```
    plt.figure(figsize=(7, 6))

    # Plot Original Pattern (Data Manifold)
    # Using 'k:' for the original circle perimeter
    plt.plot(x1[:-1], x2[:-1],'k:',label='Original Circle', alpha=.4)
    # Highlight the appended center point (the last element)
    plt.scatter(x1[-1], x2[-1], color='black', s=30, \
                                            label='Original Center')

    # Plot Linear Transformation (Affine Manifold)
    # Note: We plot all points including the transformed center
    plt.plot(Z_linear[:-1, 0], Z_linear[:-1, 1], 'r--',\
                                       label='Pure Affine (Ellipse)')
    plt.scatter(Z_linear[-1, 0], Z_linear[-1, 1], color='red', s=30,\
                                          label='Transformed Center')

    # Plot Final Activated Manifold (Model Manifold)
    plt.plot(Z_activated[:-1,0], Z_activated[:-1,1],'b-',linewidth=2,
             label=f'Activated ({activation_fn.__name__})')
    plt.scatter(Z_activated[-1, 0], Z_activated[-1, 1],\
                                                color='blue', s=30)
    plt.axhline(0, color='black', lw=0.5)
    plt.axvline(0, color='black', lw=0.5)
    plt.grid(True, linestyle=':', alpha=0.6)
    plt.axis('equal')
    plt.legend(loc='upper left', fontsize='small')
    plt.title(title)
    plt.show()
```

The following snippet plots the original circle, the purely affine-transformed circle, and the affine-transformed circle wrapped with the ReLU function.

```
x1, x2, X_bar = circle(r=1.0, dd=0.01)
Wh = W_hat.copy()   # W_hat is defined earlier
print(f"Affine transformation matrix: \n{Wh}")
plot_transformation(x1, x2, X_bar, Wh, activation_fn=relu,\
                    title="ReLU Folded Affine Transformed Circle")
```

```
Affine transformation matrix:
[[0.  0.2]
 [0.8 0.2]
 [0.2 0.5]]
```

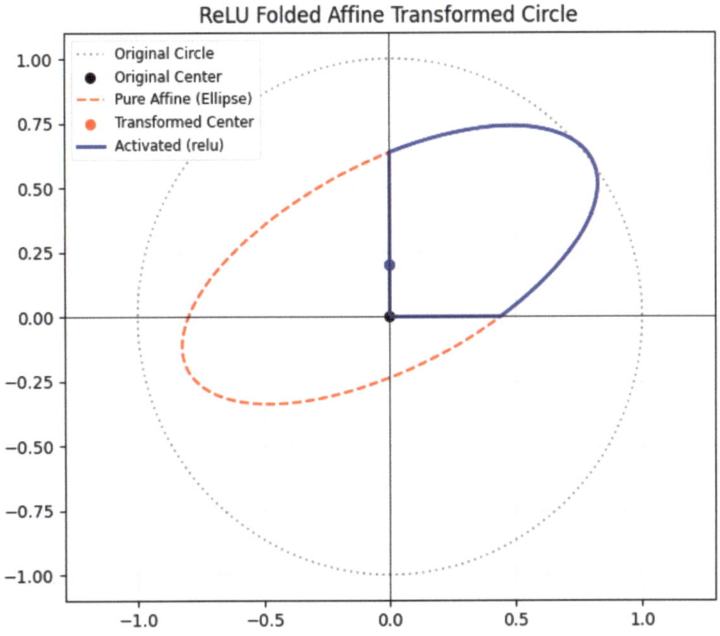

As observed, the portion of the curve that lies within the first quadrant is preserved, while segments in the other three quadrants are collapsed onto the coordinate axes due to the ReLU activation. This results in a "clipped" geometry, where any negative contributions from the affine transformation are eliminated.

It is important to note that these mappings are produced using a fixed set of predefined learning parameters. In a real machine learning model, the learning parameters are updated iteratively during training. Consequently, as the parameters change, the shape and extent of the clipped region also change. The following example illustrates this behavior.

```
Wh[0,0]=.9;  Wh[0,1]=.6      # change the biases, so no clip at all
print(f"Modified Affine transformation matrix: \n{Wh}")
plot_transformation(x1, x2, X_bar, Wh, activation_fn=relu,\
                    title="ReLU Folded Affine Transformed Circle")
```

```
Modified Affine transformation matrix:
[[0.9 0.6]
 [0.8 0.2]
 [0.2 0.5]]
```

The plot is shown in the following page.

It can be observed that the ellipse is preserved (the blue and red ellipses overlap), and the application of ReLU has no effect. Naturally, if leaving certain regions unclipped—or clipping other parts of the ellipse—helps to reduce the loss, the optimizer can achieve this by appropriately adjusting the learning parameters.

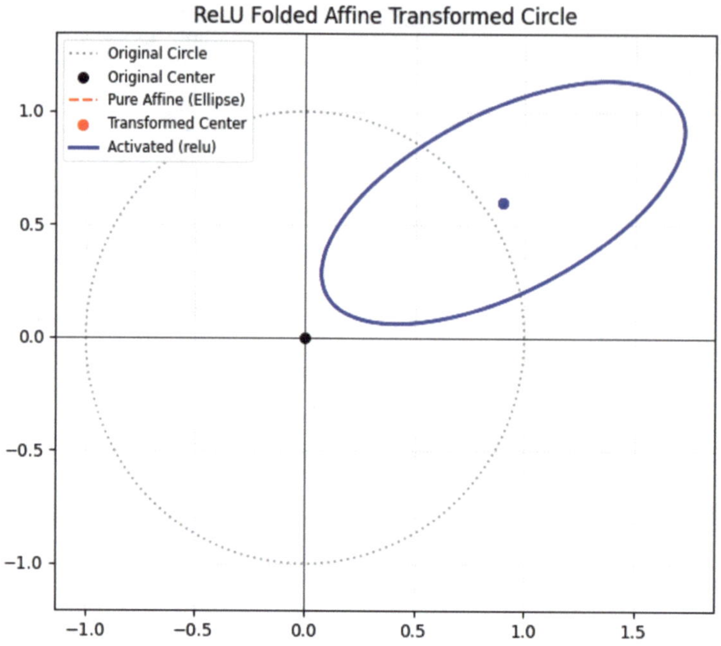

The following snippet illustrates the affine-transformed circle wrapped with the **tanh** activation function, which produces a smooth transformation:

```
x1, x2, Xb = circle(1.0,0.03)
affineplot(net_tanh, Xb, wh0_I, wh0_II, wh_I, wh_II,wh_I_b0,wh_II_b0)
```

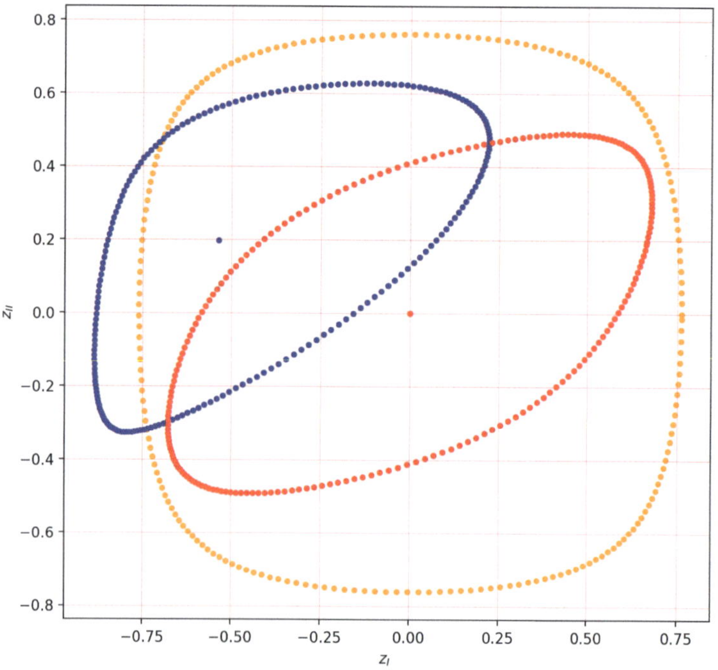

As observed, the original circle (orange), the linearly transformed circle, and the affine-transformed circle are all visibly warped under the action of **tanh**. Despite this distortion, the curves remain smooth; in fact, they become mathematically "smoother," being infinitely differentiable due to the properties of the tanh function.

In contrast to the ReLU activation, which abruptly clips the geometry and produces sharp "hinges" at the axes, **tanh** applies a continuous, nonlinear compression that preserves the topological "loop" of the circle while smoothly warping its curvature.

The intensity of the distortion induced by tanh can be adjusted by modifying the learning parameters. However, perfect preservation of the original geometry is impossible with tanh, since the equation $\tanh(z) - z = 0$ has no nonzero real solution. In contrast, with ReLU activation, exact preservation of the original shape is possible, as discussed earlier.

The following snippet plots the original circle, the purely affine-transformed circle (ellipse), and the affine-transformed circle wrapped with the **tanh** function for a direct visual comparison.

```
plot_transformation(x1, x2, X_bar, Wh, activation_fn=tanh,\
                    title="Tanh Bent Affine Transformed Circle")
```

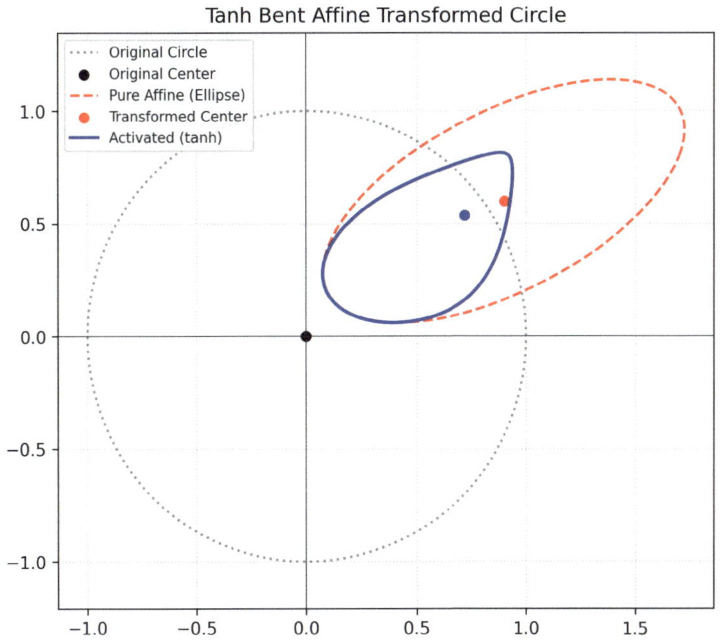

As shown, the **tanh** activation compresses the ellipses into smooth, continuous loops. The transformation preserves the continuity of the original shapes while warping their proportions to fit within the bounded range of the activation function.

9.2.4 A Spiral under Activated Affine Transformation

Unlike simple closed loops or straight lines, a spiral represents a more complex geometric structure with continuously varying curvature. When passed through a **ReLU** activation, the piecewise-linear nature of the function "breaks" the spiral wherever it crosses the coordinate axes.

The following snippet plots the original spiral, the affine-transformed spiral, and the affine-transformed spiral wrapped with the **ReLU** activation function:

```
x1, x2, Xb = spiral(0.1, 20.)
Wh = W_hat    # reset to the original W_hat
# affineplot(net_relu, Xb, wh0_I wh0_II,wh_I,wh_II,wh_I_b0,wh_II_b0)
plot_transformation(x1, x2, Xb, Wh, activation_fn=relu,\
                    title="ReLU Folded Affine Transformed Spiral")
```

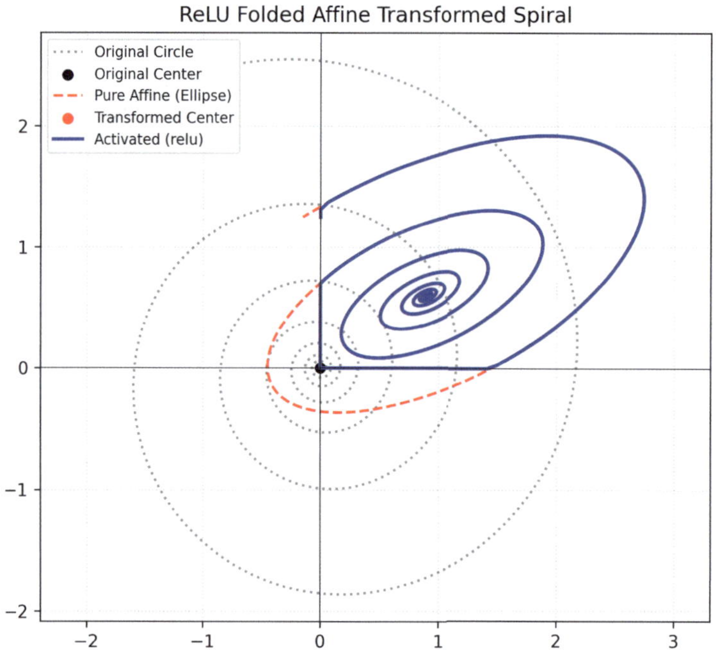

The first-quadrant preservation and the collapse of other quadrants under the action of **ReLU** are clearly observed once again. The spiral retains its original geometry wherever the affine transformation maps it into the positive domain, but it is abruptly flattened against the z-axes wherever the values become negative.

The following snippet plots the original spiral, the affine-transformed spiral, and the affine-transformed spiral wrapped with the **tanh** activation function:

```
# x1, x2, Xb = spiral(0.1,10.)
```

```
# affineplot(net_tanh, Xb, wh0_I, wh0_II, wh_I,␣
↪wh_II,wh_I_b0,wh_II_b0)
plot_transformation(x1, x2, Xb, Wh, activation_fn=tanh,\
                    title="Tanh Bent Affine Transformed Spiral")
```

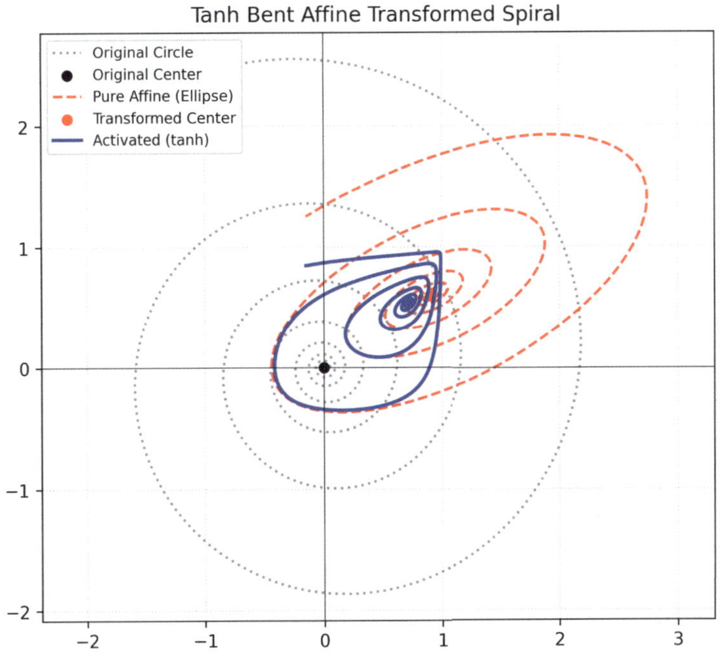

As shown, the **tanh** activation compresses the spiral into smooth, curved loops. In this transformed state, it no longer resembles a traditional spiral: the infinite expansion is curtailed, and the entire structure is confined within the range $(-1, 1)$ along both dimensions.

This bounding effect is characteristic of saturating activation functions. While the affine transformation can push values toward infinity, the **tanh** function maps the entire real line into a finite, open interval. As a result, the outer layers of the spiral are increasingly compressed toward the boundaries of the square defined by the co-domain, all while preserving a smooth, continuous, and differentiable manifold.

Chapter 10

Concept of Parameter Encoding

Once a model is trained, the training dataset is no longer strictly required because its information has been captured within the model's learning parameters. This idea of **parameter encoding** provides a clear framework for understanding how learning occurs through affine transformations.

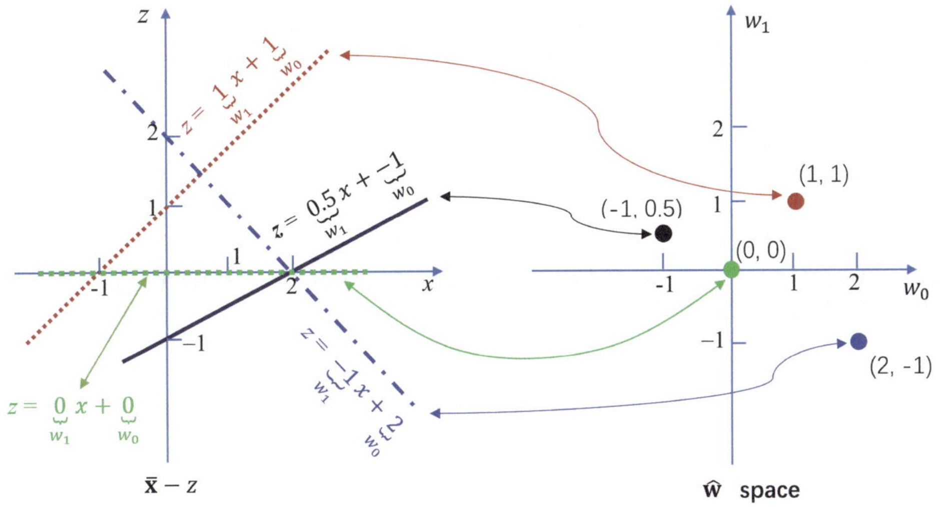

Figure 10.1: Data representing the relationship between $\bar{\mathbf{x}}$ and z (or y) is encoded into model parameters $\hat{\mathbf{w}}$ in the hypothesis space. Effectively, a machine learning model converts raw data into $\hat{\mathbf{w}}$ during training. Prediction is then performed by using those parameters $\hat{\mathbf{w}}$ to estimate z for new features within the same hypothesis.

10.1 $\overline{\mathbf{X}}$ to $\hat{\mathbf{w}}$ Encoding: The Data–Parameter Converter

Based on Eq. (2.4), we can view a dataset $(\bar{\mathbf{x}}, z)$ as being encoded into a set of learning parameters $\hat{\mathbf{w}}$ within the hypothesis space where all parameters reside. Figure 10.1

schematically illustrates this encoded state.

As illustrated, every straight line in the input space is represented by a single point in the hypothesis space $\mathbb{R}^2$. For example, the red line is encoded by a red dot at $w_0 = 1$ and $w_1 = 1$, which allows the model to exactly reproduce that line. Similarly, the blue line is encoded by a blue dot at $w_0 = 2$ and $w_1 = -1$.

The machine learning process aims to identify the optimal coordinates (dots) in this hypothesis space based on the provided dataset. Once these parameters are determined, the model can generate the corresponding lines (functions). In essence, the model acts as a **converter** that encodes data into w_i. Consequently, the scale and quality of the dataset directly influence the required dimensions of the affine subspaces within the model. While modern machine learning models are far more complex, this encoding process represents their fundamental mechanism at the level of individual neurons.

10.2 Uniqueness of the Encoding

We assert that the encoding of a line in the $\overline{\mathbf{x}}$–z space as a point in the hypothesis space is unique. This claim can be established directly as follows.

Assume that an arbitrary line in the $\overline{\mathbf{x}}$–z space corresponds to two distinct points in the hypothesis space, denoted by $\hat{\mathbf{w}}^{(1)}$ and $\hat{\mathbf{w}}^{(2)}$. According to Eq. (4.1), this line may be written as

$$z = \overline{\mathbf{x}}\hat{\mathbf{w}}^{(1)} \tag{10.1}$$

Because this expression must hold for any arbitrary $\overline{\mathbf{x}}$, and because the same line is also described by

$$z = \overline{\mathbf{x}}\hat{\mathbf{w}}^{(2)} \tag{10.2}$$

subtracting Eq. (10.1) from Eq. (10.2) yields

$$0 = \overline{\mathbf{x}}\left[\hat{\mathbf{w}}^{(2)} - \hat{\mathbf{w}}^{(1)}\right] \tag{10.3}$$

Since this equality must be satisfied for all values of $\overline{\mathbf{x}}$, the quantity in brackets must vanish. Therefore,

$$\hat{\mathbf{w}}^{(1)} = \hat{\mathbf{w}}^{(2)} \tag{10.4}$$

This establishes uniqueness in one direction.

Conversely, we show that a single point in the hypothesis space defines a unique line in the $\overline{\mathbf{x}}$–z space.

Let an arbitrary point $\hat{\mathbf{w}}$ in the hypothesis space be given, and suppose that it generates two distinct lines, $z^{(1)}$ and $z^{(2)}$. From Eq. (4.1), these lines satisfy

$$z^{(1)} = \overline{\mathbf{x}}\hat{\mathbf{w}} \tag{10.5}$$

$$z^{(2)} = \overline{\mathbf{x}}\hat{\mathbf{w}} \tag{10.6}$$

Subtracting the two expressions gives

$$z^{(2)} - z^{(1)} = 0 \;\Rightarrow\; z^{(2)} = z^{(1)} \tag{10.7}$$

which shows that the two lines are identical. This completes the proof of uniqueness.

Geometrically, this one-to-one correspondence is evident in Fig. 10.1: a line is uniquely determined by its slope and bias, both of which are explicitly encoded in the components of $\hat{\mathbf{w}}$.

This uniqueness property is a fundamental reason why high-quality datasets can be reliably encoded into learning parameters via affine transformations. It ensures that a machine learning model can represent and learn from the underlying structure of the data in a consistent and unambiguous manner.

Finally, we note that when an affine transformation is composed with an activation function, this uniqueness is preserved provided the activation function is **strictly monotonic**. This condition is satisfied by all standard activation functions discussed in Chapter 7 of Ref. [9].

10.3 On Non-monotonic Activations

Although uniqueness holds for strictly monotonic activation functions, it is important to recognize that non-monotonic activations—such as the sine function employed in Sinusoidal Representation Networks (SIREN)—violate this one-to-one correspondence.

In a monotonic system, the mapping between the hypothesis space and the label space is bijective: each point in the parameter space $\hat{\mathbf{w}}$ defines exactly one prediction function. In contrast, a periodic function such as $\sin(z)$ is inherently many-to-one. Because $\sin(z) = \sin(z + 2\pi n)$, multiple distinct parameter vectors $\hat{\mathbf{w}}$ can produce identical prediction outputs z for the same input $\overline{\mathbf{x}}$.

This form of mathematical "aliasing" has several important implications for the encoding concept:

- **Redundancy in Hypothesis Space:** A dataset is no longer represented by a unique point but by a set of equivalent points in the parameter space.
- **Optimization Landscape:** The loss surface becomes more complex and rugged, containing multiple equivalent global or local minima.
- **Representational Power:** Despite the loss of uniqueness, non-monotonic activations are often deliberately employed to enhance the network's ability to represent high-frequency features and complex signals that monotonic functions cannot efficiently capture.

Consequently, the reliability and interpretability of the parameter encoding discussed in this section are most rigorously defined when the activation function σ is strictly monotonic.

10.4 The Gradient of the Prediction Function

10.4.1 Gradient of Prediction Function

The gradient of the prediction function used in the affine transformation with respect to the learning parameters takes the following simple forms:

$$\begin{aligned} \nabla_{\mathbf{w}} z &= \nabla_{\mathbf{w}}(b + \mathbf{x}\mathbf{w}) = \mathbf{x} \\ \nabla_b z &= \nabla_b(b + \mathbf{x}\mathbf{w}) = 1 \end{aligned} \tag{10.8}$$

These expressions reveal that the gradient with respect to the weights is identical to the feature variables (the data). In contrast, the gradient with respect to the bias is a constant unit. This is mathematically consistent because the "feature" corresponding to the bias is effectively $x_0 = 1$. This distinction suggests that when applying regularization techniques (see Chapter 13 of Ref. [9]), one may choose different regularization parameters for weights and bias to account for their differing roles.

When the compact $\overline{\mathbf{x}}\hat{\mathbf{w}}$ form is used, the relationship becomes:

$$\nabla_{\hat{\mathbf{w}}} z = \nabla_{\hat{\mathbf{w}}}(\overline{\mathbf{x}}\hat{\mathbf{w}}) = \overline{\mathbf{x}} \tag{10.9}$$

This elegant form is what is typically utilized by autograd engines during the optimization phase of machine learning processes. It emphasizes that the direction and magnitude of the steepest update in the parameter space are dictated by the input data itself. Consequently, the quality of the dataset—including its scale, noise levels, and outliers—directly governs the stability and trajectory of the learning process.

10.4.2 Gradient of the Wrapped Prediction Function

In modern machine learning, the affine transformation is typically "wrapped" in a non-linear activation function $\phi(\cdot)$ to form the final prediction $a = \phi(z)$. To find the gradient of this wrapped function with respect to the parameters, we apply the chain rule.

Let the wrapped prediction be $\hat{y} = \phi(\bar{\mathbf{x}}\hat{\mathbf{w}})$. The gradient with respect to the augmented parameters $\hat{\mathbf{w}}$ is given by:

$$\nabla_{\hat{\mathbf{w}}}\hat{y} = \frac{d\hat{y}}{dz}\nabla_{\hat{\mathbf{w}}}z = \phi'(z)\bar{\mathbf{x}} \tag{10.10}$$

Expanding this back into separate weight and bias components:

$$\begin{aligned} \nabla_{\mathbf{w}}\hat{y} &= \phi'(b + \mathbf{x}\mathbf{w})\mathbf{x} \\ \nabla_{b}\hat{y} &= \phi'(b + \mathbf{x}\mathbf{w}) \end{aligned} \tag{10.11}$$

In these formulas, $\phi'(z)$ acts as a "scaling factor" or "gate" for the data. If the activation function is in a saturated region (where the slope $\phi'(z) \approx 0$), the gradient effectively disappears—a phenomenon known as the vanishing gradient problem. This further illustrates that in a wrapped model, the parameter update is a product of both the input data $\mathbf{x}$ and the current state of the model's activation.

This derivation implies that the activation function must be differentiable, if a gradient based algorithm is used to train the model.

Chapter 11

Transformation Using High-Order Basis

As an extension of the standard affine transformation, an arbitrary function can be approximated by employing a set of linearly independent nonlinear basis functions, such as monomials [10]. Consider an arbitrary function $y(\mathbf{x})$ that can be expressed as a linear combination of known basis functions:

$$y(\mathbf{x}) = \overline{\mathbf{x}}_{\text{monomials}}\, \hat{\mathbf{k}} \tag{11.1}$$

where $\overline{\mathbf{x}}_{\text{monomials}}$ is the row vector of monomial basis functions:

$$\overline{\mathbf{x}}_{\text{monomials}} = [1, x_1, \dots, x_p, x_1 x_2, x_1 x_3, \dots, x_1 x_p, x_1^2, x_2^2, \dots] \tag{11.2}$$

Assuming there are p' terms in $\overline{\mathbf{x}}_{\text{monomials}}$, the corresponding coefficient vector $\hat{\mathbf{k}}$ is defined as:

$$\hat{\mathbf{k}} = [k_0, k_1, \dots, k_{p'}]^\top \tag{11.3}$$

The prediction function z is then defined similarly:

$$z = \overline{\mathbf{x}}_{\text{monomials}}\, \hat{\mathbf{w}}^* \tag{11.4}$$

By setting the weight vector $\hat{\mathbf{w}}^* = \hat{\mathbf{k}}$, the prediction becomes exact, demonstrating the model's ability to reproduce the underlying function.

In this nonlinear case, the transformation is no longer affine. This approach is rarely implemented directly in deep neural network models because the basis dimensionality expands rapidly as the order of the monomials increases—a phenomenon known as the "curse of dimensionality." To illustrate this, consider the growth of the number of terms (p') for a set of p input features expanded to a polynomial degree d. The total number of basis functions (including the bias/constant term) is determined by the following combination formula:

$$\binom{p+d}{d} = \frac{(p+d)!}{p!d!}$$

In scenarios where only feature interactions are required, we may limit the expansion to **bilinear terms**. This excludes pure power terms (e.g., x_1^2), focusing solely on the products of distinct feature pairs $x_i x_j$ where $i \neq j$. In this case, the number of basis functions is reduced to:

$$p' = 1 + p + \binom{p}{2}$$

The table below compares the growth of p' for full quadratic expansions versus bilinear-only expansions:

Features (p)	Expansion Type	Formula	Total Terms (p')
2	Full Quadratic ($d = 2$)	$\binom{2+2}{2}$	6
2	Bilinear Only	$1 + 2 + \binom{2}{2}$	4
3	Full Quadratic ($d = 2$)	$\binom{3+2}{2}$	10
3	Bilinear Only	$1 + 3 + \binom{3}{2}$	7
10	Full Quadratic ($d = 2$)	$\binom{10+2}{2}$	66
10	Bilinear Only	$1 + 10 + \binom{10}{2}$	56
100	Full Quadratic ($d = 2$)	$\binom{100+2}{2}$	5,151
100	Bilinear Only	$1 + 100 + \binom{100}{2}$	5,051

While direct expansion is often impractical for high-dimensional neural networks, higher-order or nonlinear basis functions (such as Radial Basis Functions, or RBFs) remain a staple of nonlinear regression. More advanced machine learning models utilize the "kernel trick" to navigate these high-dimensional spaces efficiently—a technique particularly effective in Support Vector Machines (SVMs) [2, 9].

This reproducibility property is analogous to the use of higher-order elements in the Finite Element Method (FEM) [18], where increased polynomial orders improve approximation accuracy within an element. This practice is highly effective in FEM because the spatial dimensionality (the number of independent variables) is typically limited to 3, preventing the combinatorial explosion seen in high-dimensional feature spaces.

Chapter 12

Affine Manifold

Briefly stated:

> A **manifold** is a topological space that locally resembles Euclidean space near each point.

12.1 Geometric Learning Framework

In the modern geometric deep learning framework, we operate under the **manifold assumption**:

> High-dimensional data often lie near a low-dimensional manifold embedded in the ambient space. The observed data are samples from (or noisy observations of) that manifold.

In many real-world problems, datasets exhibit the following structural characteristics:

- **Local Simplicity:** Each small neighborhood is simple and approximately linear.
- **Global Complexity:** The overall global structure is non-linear and complex.
- **Emergent Structure:** While raw input feature spaces may not be geometrically well-structured, transformations within a model allow learned representations to organize data onto geometric manifolds.

Counterexamples to the Manifold Assumption

The manifold assumption may fail in the following cases:

- **Independent High-Dimensional Features:** When feature variables are completely independent, a lower-dimensional manifold cannot emerge.
- **Discrete/Combinatorial Features:** In cases where features are discrete or combinatorial, local linearity is significantly weakened.

- **Dominant Noise:** Strong noise acts as a "blur" on the data manifold. If data quality is poor, the "true" surface remains hidden behind a cloud of points.

For datasets that align with the manifold assumption, training a machine learning model is the process of constructing and fitting a **model manifold** to the **data manifold**, as shown in Fig. 12.1.

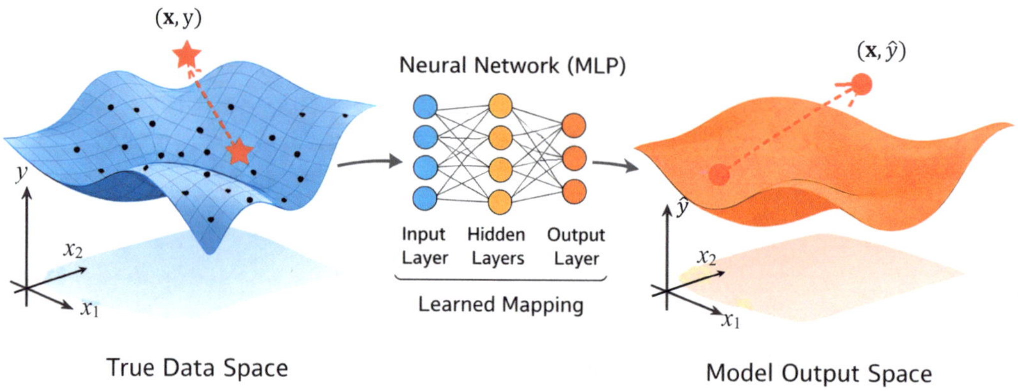

Figure 12.1: A schematic geometric interpretation of supervised regression with a Multi-Layer Perceptron (MLP). Image adapted from a conceptual diagram generated by ChatGPT.

Fig. 12.1 illustrates the relationship between the true data manifold and the model manifold in a shared joint feature–label space. The blue surface (2D) in the 3D ambient space represents the true data manifold defined by the graph of an unknown function $\mathbf{y}(\mathbf{x})$, where black dots denote finite, possibly noisy samples. The orange surface represents the model manifold defined by the learned ML model, such as an MLP function $\hat{\mathbf{y}}(\mathbf{x})$. Geometrically, training the MLP involves adjusting learning parameters so the model manifold aligns as closely as possible with the true data manifold.

An MLP model manifold is constructed from a large number of local affine mappings (locally linear) connected seamlessly by activation functions (enabling global complexity). In this context, activation functions "glue" together the affine mappings across all neurons and layers.

- **ReLU Activation:** Acts like a hinge, folding or unfolding affine mappings. The resulting manifold surface is continuous, but derivatives may be discontinuous at the hinge points.
- **Tanh/Sigmoid Activation:** Results in a model manifold that is smooth and theoretically infinitely differentiable.

We thus observe:

A neural network model excels at fitting datasets with manifold structures.

It forms an **affine manifold** where affine mappings provide local linearity and activation functions provide global geometric complexity.

Developing a regression model is essentially creating a continuous surface through a specialized form of **interpolation** (affine and activation-based) using discrete data points. This is perhaps the **closest analogy** to conventional computational methods like FEM, S-FEM, or meshfree methods, which perform interpolations over a physical structure, albeit with strict stability controls such as the energy principle [18] or weakened weak forms [5, 17].

12.2 Example: Regression as a 3D Curve Fit

12.2.1 Setting of the Problem

If a dataset has p input feature variables and we are predicting a continuous function (or a vector of k target values), we are essentially asserting that the targets lie on a surface or curve defined by:

$$\hat{\mathbf{y}} = f(\mathbf{x}) \tag{12.1}$$

- **The Data Manifold:** This is the underlying "true" relationship between input features and output targets.
- **The Model Manifold:** An MLP creates a continuous, multidimensional "sheet." Training is the process of pulling and stretching this sheet—by adjusting weights and biases—until it aligns as closely as possible with the training data points.
- **Example Setting:** Consider inputting "Time" ($p = 1$) to predict the (x, y, z) coordinates ($k = 3$) of a moving object. In this case, the model manifold is a **1D curve** snaking through a **3D space**. The MLP's task is to learn the geometrical path that defines this manifold.

We assume the true label functions are:

$$\mathbf{y} = \begin{bmatrix} \sin x \\ \cos x \end{bmatrix} \tag{12.2}$$

The **dataset** is generated using these functions. We first create a NumPy array for x in the domain $(-1, 1)$, substitute these values into Eq.(12.2) to compute the two label arrays, and then combine them into a 2D array.

12.2.2 Build and Train the Regression Manifold

In this example, we map a single input x ($p = 1$) to two outputs (y_1, y_2) ($k = 2$) using three hidden layers. The hidden layer architecture is `(10, 50, 10)`. The complete sequence of layers in the MLP is therefore `(1, 10, 50, 10, 2)`.

When accounting for the bias terms (equivalent to adding a neuron with a constant value of 1 to each layer except the output), the effective layer dimensions become (2, 11, 51, 11, 2). This results in a total of 75 neurons involved in affine transformations, giving the MLP a total pseudo-dimension of $P^* = 75$. The total number of learning parameters available to shape the affine manifold is $P = 1102$ $(2 \times 10 + 11 \times 50 + 51 \times 10 + 11 \times 2)$ [9]. Clearly, the Neurons-Samples Theorem, Eq.(4.13), is satisfied.

For this demonstration, we use the **ReLU activation** function for all hidden layers, though this can easily be swapped for Tanh or others. Geometrically, the resulting MLP model manifold represents a **1D curve** snaking through **3D space**, where the outputs $\hat{\mathbf{y}}$ form the coordinate axes.

We utilize the `scikit-learn` module to build and train the MLP. The following code demonstrates how to implement this using synthetic data:

```
from sklearn.neural_network import MLPRegressor
from mpl_toolkits.mplot3d import Axes3D
# 1. Generate Data: one feature variable and two target functions:
# y = [sin(x), cos(x)],  p=1, k=2

m = 500             # number of samples (measurements)
X = np.linspace(0, 2 * np.pi, m).reshape(-1, 1)                  # p=1
y1 = np.sin(X).ravel() + np.random.normal(0, 0.05, m)           # k=2
y2 = np.cos(X).ravel() + np.random.normal(0, 0.05, m)
Y = np.column_stack((y1, y2))

# 2. Create a MLP model with 3 hidden layers; or activation="tanh"
mlp = MLPRegressor(hidden_layer_sizes=(10, 50, 10), max_iter=1000, \
                              activation="relu", random_state=42)
mlp.fit(X, Y) # fit the MLP manifold
# Count total number of learing parameters
n_params = sum(w.size for w in mlp.coefs_) + \
           sum(b.size for b in mlp.intercepts_)
print(f"Total number of learning parameters used ")
print(f"      in all the affine mappings in MLP: {n_params}")

# 3. Predict using the trained MLP manifold
X_test = np.linspace(0, 2 * np.pi, 300).reshape(-1, 1)
Y_hat = mlp.predict(X_test)
# Visualizing the Affine Regression Manifold
# The following snippet plots affine MLP regression manifold in 3D.
# Optimized 3D Visualization
fig = plt.figure(figsize=(10, 7))
ax = fig.add_subplot(111, projection='3d')
```

```
# Scatter and Line Plot
ax.scatter(X.ravel(), Y[:, 0], Y[:, 1], c=X.ravel(), cmap='jet',\
                                alpha=0.4, label='Data Samples')
ax.plot(X_test.ravel(), Y_hat[:, 0], Y_hat[:, 1], color='red',\
                                 lw=4, label='Learned Manifold')
# Adjusting labels and padding to prevent clipping
ax.set_xlabel('Input (x)', labelpad=5)
ax.set_ylabel('Output 1 (y1)', labelpad=5)
ax.set_zlabel('Output 2 (y2)', labelpad=2)
ax.set_title('Manifold Mapping (p=1 to k=2)', pad=-15)

# Adjust the viewing angle so labels are more visible
ax.view_init(elev=45, azim=-15)
# Tight layout ensures the plot fits within the figure canvas
plt.tight_layout()
plt.legend(loc='center left', bbox_to_anchor=(0.2, 0.75))
plt.show()
```

```
Total number of learning parameters used
      in all the affine mappings in MLP: 1102
```

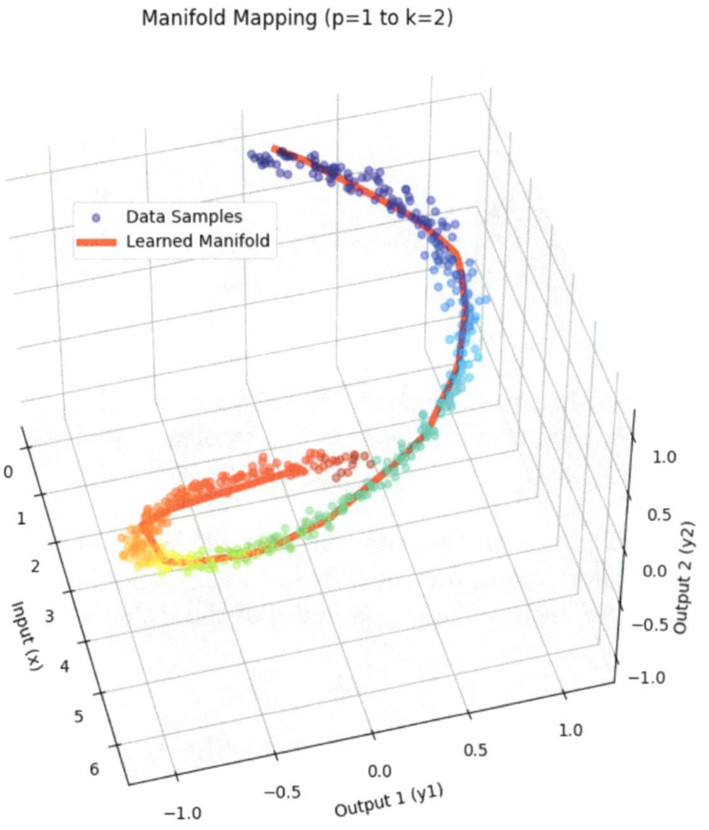

The foregoing plot illustrates the model manifold learned through an affine MLP regression of a curve in 3D space. Because the ReLU activation function is employed for each neuron, the characteristic piecewise linear (affine) structure is clearly visible. The manifold appears as a series of connected linear segments, reflecting the "hinge" behavior of the ReLU units.

In contrast, if a tanh (smooth) activation is used, the resulting manifold is significantly different. As shown in Fig. 12.2, the tanh activation produces a smooth, continuously differentiable manifold that curves gracefully through the space.

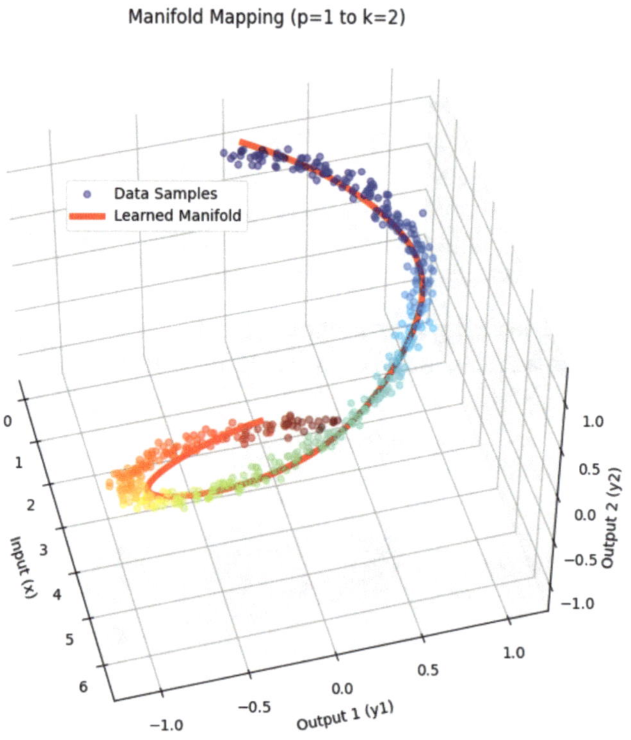

Figure 12.2: An affine MLP regression manifold of a curve in 3D space using the tanh activation function for each neuron, resulting in a smooth, differentiable manifold.

This comparison highlights a fundamental property of the affine manifold: while the underlying affine transformations provide the local scaffolding, the choice of activation function determines the "texture" and continuity of the global geometric representation.

12.3 Observations from the Example

Hinged Flat Affine Mappings

As illustrated in the preceding plots, the MLP manifold is composed of a sequence of flat affine mappings (straight line segments) propagated through successive layers of neurons and connected by nonlinear activation functions. This structure is especially apparent when the **ReLU** activation is employed, revealing the explicitly **piecewise linear** character of the manifold. In this regime, the learned manifold resembles a form of "high-dimensional origami," where the ReLU activation acts as a hinge that folds or unfolds adjacent affine mappings. By contrast, when **tanh** is used, these affine segments are bent smoothly rather than folded sharply.

Affine Mappings as Building Blocks

In each neuron, the transformation applied prior to the activation function is an **affine mapping**, as defined in Eq. (4.7). From a geometric perspective, an affine map can rotate, scale, shear, and translate line segments. Importantly, it cannot bend them; the segments remain **straight** under the transformation.

Activation as Glue. It Joints and Bends in Affine Mappings

The nonlinear activation function (e.g., $\mathrm{ReLU}(z) = \max(0, z)$) functions as a **fold** or **hinge**. On one side of the activation threshold ($z > 0$), the mapping remains a flat affine identity. On the other side ($z < 0$), the mapping is collapsed to zero, producing a sharp geometric transition.

The Result: A Piecewise Affine Manifold

By stacking multiple layers of straight affine mappings separated by activation-induced folds, the MLP realizes a **continuous piecewise affine** function. The resulting model manifold consists of numerous linear "segments" or "regions" joined together at varying angles, collectively forming a complex global geometry.

12.3.1 Summary: From Local Flatness to Global Curvature

Component	Mathematical Operation	Geometric Effect
Weights & Biases	Affine Transformation ($\overline{\mathbf{x}}\widehat{\mathbf{w}}$)	Stretches, rotates, shears, and shifts locally flat space. **Brick-like**
Activation (ReLU)	Nonlinear Gate	Introduces a "fold" or "crease" in the manifold. **Glue-like**
Deep Architecture	Function Composition	Aggregates many local folds to produce an apparently curved geometry.

Readers may experiment with replacing the `relu` activation with `logistic` in the provided code. The resulting manifold appears significantly smoother because the `logistic` function is C^∞ continuous (as is `tanh`). These activations bend affine mappings while simultaneously smoothing transitions between regions. The key differences are summarized below.

12.3.2 ReLU Manifold vs. Tanh/Logistic Manifold

Feature	ReLU MLP Manifold	Tanh or Logistic MLP Manifold
Local Geometry	Flat (Affine facets/lines)	Curved (Nonlinear)
Transitions	Sharp "hinges" (creases)	Smooth "bends"
Differentiability	Not differentiable at hinges	Infinitely differentiable
Visual Analogy	**Origami-like:** many flat paper folds	**Clay-like:** a smoothly molded surface

12.4 Relationship Between Network Width and Manifold Complexity

The geometric complexity of an affine manifold is directly constrained by the number of neurons in the hidden layers. In a ReLU-based network, each neuron in a hidden layer defines a hyperplane that can introduce a local "crease" or "fold" in the manifold.

- **Width as "Fold Capacity":** The number of neurons (i.e., the width) of a layer determines the maximum number of linear regions the model can generate. A wider layer provides more "hinges," enabling the manifold to bend around complex data structures or to approximate curved geometries with higher local fidelity.
- **Depth as "Fold Interaction":** While increasing width introduces additional creases, increasing depth (i.e., stacking more layers) allows these creases to interact and fold upon one another. This interaction leads to an exponential growth in the number of linear regions, giving rise to the "high-dimensional origami" effect discussed earlier.

Architectural Change	Geometric Result
Increasing Width	Adds more "facets" or "segments" to the manifold, improving local resolution.
Increasing Depth	Enables complex global folding, allowing the representation of highly oscillatory or nested functions.

For example, in the previous 1D-to-2D mapping, a hidden layer with only 2 neurons can produce at most 3 linear segments. By increasing the width of the middle layer to 50 neurons, the model is endowed with a sufficient number of "hinges" to approximate the smooth transcendental curves of $\sin(x)$ and $\cos(x)$ so closely that the underlying piecewise-linear structure becomes nearly imperceptible to the naked eye.

12.5 Example: Regression as a 3D Surface Fit

12.5.1 Setting of the Problem

This example demonstrates the generation of a **curved surface**—a two-dimensional (2D) manifold embedded in three-dimensional (3D) space—so that the learned model can be conveniently plotted and visualized. To achieve this, the MLP architecture must represent a dimension-reduced $2 \to 1$ mapping, meaning that there are **two input features** ($p = 2$) and **one target output** ($k = 1$).

Although the visualization space remains 3D, the learned manifold becomes a surface rather than a curve, because the model now depends on two independent variables. For this demonstration, we use the following simple function as the true label function, defined over a square domain:

$$y = \sin \sqrt{x_1^2 + x_2^2} \tag{12.3}$$

The **dataset** is generated by evaluating this function on a 2D grid. The code used in the previous example can be extended to accomplish this task. The key changes are as follows:

1. The input dimension is increased to 2, and a coordinate grid (`np.meshgrid`) is used to define the input space, allowing the MLP to learn a function of two variables.
2. The output dimension is reduced to 1, which is visualized as a 2D surface embedded in 3D space.
3. The learned manifold is plotted using `ax.plot_surface`, which visualizes the continuous 2D surface that the neural network has effectively "bent" through the 3D space to fit the data points.

12.5.2 Build and Train the Regression Manifold

We utilize the `scikit-learn` module to build and train the MLP. The following code demonstrates how to create a mesh grid, and generate a synthetic dataset using the function defined in Eq.(12.3).

```
from sklearn.neural_network import MLPRegressor

# 1. Generate Data: p=2 (inputs), k=1 (output)
m_side = 30  # number grid on each side of the square
x_start = -3; x_end = 1
x1 = np.linspace(x_start, x_end, m_side)
x2 = np.linspace(x_start, x_end, m_side)
X1, X2 = np.meshgrid(x1, x2)
```

```
# Flatten the grid to create (900, 2) input matrix
X_grid = np.column_stack((X1.ravel(), X2.ravel()))

# Calculate Y targets based on the flattened R
# Ensure Y matches the number of samples in X_grid (900)
R = np.sqrt(X_grid[:, 0]**2 + X_grid[:, 1]**2)
# Add in some noise:
Y = np.sin(R) + np.random.normal(0, 0.02, len(R)) # Shape is (900,)
```

12.5.3 Create and Fit MLP

We use 'relu' as the activation function to fold the manifold:

```
mlp = MLPRegressor(hidden_layer_sizes=(10, 50, 10), max_iter=2000,
                   activation="relu", random_state=42)  # or tanh
mlp.fit(X_grid, Y)
# 3. Predict for the Manifold Visualization
x_test_side = 50
x1_t = np.linspace(x_start, x_end, x_test_side)
x2_t = np.linspace(x_start, x_end, x_test_side)
X1_t, X2_t = np.meshgrid(x1_t, x2_t)
X_test_flat = np.column_stack((X1_t.ravel(), X2_t.ravel()))
Y_hat = mlp.predict(X_test_flat)
Y_hat_surface = Y_hat.reshape(x_test_side, x_test_side)

# 4. Visualization
fig = plt.figure(figsize=(12, 8),dpi=300)
ax = fig.add_subplot(111, projection='3d')
# Scatter the original noisy samples  'viridis', 'jet', 'plasma'
ax.scatter(X_grid[:, 0], X_grid[:, 1], Y, c=Y, cmap='viridis',
           alpha=0.3, label='Training Data')

# Plot the learned MLP surface
surf = ax.plot_surface(X1_t, X2_t, Y_hat_surface, cmap='viridis',
                                 alpha=0.8, antialiased=True)
ax.set_xlabel('Input X1')
ax.set_ylabel('Input X2')
ax.set_zlabel('Predicted Y')
ax.set_title('MLP Learned Surface Manifold (p=2 -> k=1)')
fig.colorbar(surf, shrink=0.5, aspect=10)
plt.show()
```

The following shows an affine MLP regression manifold of a surface in 3D space trained using the ReLU activation function for each neuron:

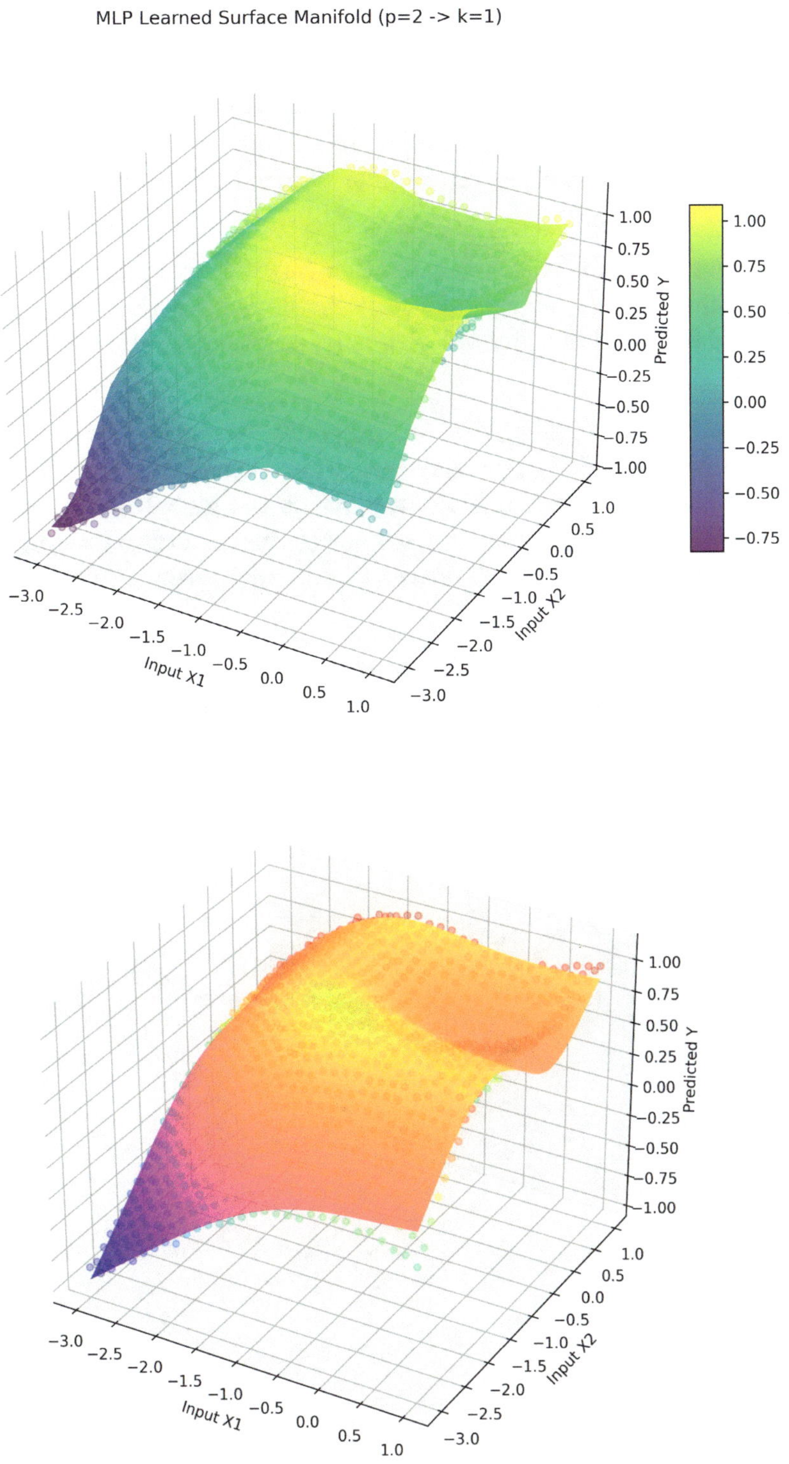

Figure 12.3: An affine MLP regression manifold of a surface in 3D space using the tanh activation function for each neuron, resulting in a smooth and differentiable manifold.

12.5.4 Folded Plane vs. Bent Plane

As in the previous example of a curved manifold in 3D space, the characteristic piecewise planar (affine) structure is clearly visible when the ReLU activation function is employed in each neuron. The resulting manifold appears as a collection of connected linear plane segments, reflecting the familiar "hinge-line" behavior induced by ReLU units.

In contrast, when a tanh (smooth) activation function is used, the learned manifold is markedly different. As shown in Fig.12.3, the tanh activation yields a smooth, continuously differentiable manifold that curves gracefully through the embedding space.

This comparison again highlights a fundamental geometric property of affine MLP manifolds: while the underlying affine transformations provide the local linear scaffolding, the choice of activation function governs the global smoothness, continuity, and overall "texture" of the resulting geometric representation. Affine MLP manifolds can be viewed as topological spaces that locally resemble Euclidean space in a neighborhood of each point.

In general, an affine MLP manifold can be successfully trained for a given problem provided that the problem is well posed and that the available data are sufficient in quantity and of good quality. The same holds when the data are generated not by a smooth function but by a partial differential equation (PDE), as long as the PDE is well posed in the Hadamard sense. Physics-informed neural networks provide a representative example of this setting [12, 19].

Chapter 13

Comparison between MLP and FEM

Although the linear prediction function and the affine transformation share the same mathematical foundation—both being defined by the hypothesis in Eq. (2.4)—they represent **distinct operational paradigms** in computational science and machine learning.

13.1 Nodal Interpolation vs. Manifold Interpolation

In **Finite Element Methods (FEM)** and related frameworks such as S-FEM [17] or meshfree methods [5], the primary unknowns are **nodal displacements** (e.g., u, v, w in solid mechanics). The physical domain is discretized into a fixed mesh, and the linear prediction function is used to approximate displacement fields within each element. Inter-element continuity is enforced through shape functions, yielding a **piecewise linear** displacement field that is constrained by boundary conditions and governed by physical (mechanics, energy principles) laws.

In contrast, a **Multi-Layer Perceptron (MLP)** employs affine transformations as fundamental building blocks to construct a **piecewise linear** manifold. In this setting, the variables are abstract **learning parameters** $(\mathbf{w}, b)$, which—together with the activation functions—control the geometry of the learned manifold. Rather than relying on linear finite elements, an MLP consists of flat affine planes. Nonlinear activation functions (such as ReLU) act to "glue" these affine planes together, producing a **piecewise affine** manifold. Unlike FEM, this manifold is not fixed in physical space; instead, it is learned and continuously reshaped to fit the underlying data distribution.

13.2 Summary of Conceptual Differences

Aspect	Finite Element Method (FEM)	Multi-Layer Perceptron (MLP)
Primary Variable	Nodal Displacements (Physical)	Weights and Biases (Learning)
Connectivity	Shape Function Continuity (C^0, C^1, etc.)	Nonlinear Activation (ReLU C^0, Tanh C^∞)
Geometric Form	Piecewise Linear Field	Piecewise Affine Manifold
Optimization	Energy Principles Governed by Physical Laws	Loss Minimization Driven by Data
Domain	Fixed Mesh / Physical Space	Abstract Feature Space

In essence, FEM employs **linear segments** to approximate a physical displacement field over a predefined spatial domain, whereas an MLP uses **affine segments** to interpolate and construct an emergent geometric manifold directly from discrete data.

Bibliography

[1] Tom M. Apostol. *Mathematical Analysis.* Addison-Wesley, Reading, MA, 2 edition, 1974.

[2] Bernhard E. Boser, Isabelle M. Guyon, and Vladimir N. Vapnik. Training algorithm for optimal margin classifiers. In *Proceedings of the Fifth Annual Workshop on Computational Learning Theory (COLT '92)*, Pittsburgh, PA, USA, 1992. ACM. Introduces nonlinear SVMs via the kernel trick.

[3] Kaiming He, Xiangyu Zhang, Shaoqing Ren, and Jian Sun. Deep residual learning for image recognition. *CoRR*, abs/1512.03385, 2015.

[4] Kaiming He, Xiangyu Zhang, Shaoqing Ren, and Jian Sun. Delving deep into rectifiers: Surpassing human-level performance on imagenet classification. In *Proceedings of the IEEE International Conference on Computer Vision (ICCV)*, pages 1026–1034, 2015.

[5] G.R. Liu. *Mesh free methods: moving beyond the finite element method, 2nd Ed.* CRC press, Taylor and Francis Group, New York, 2010.

[6] G.R. Liu. FEA-AI and AI-AI: Two-way deepnets for real-time computations for both forward and inverse mechanics problems. *International Journal of Computational Methods*, 16(08):1950045, 2019.

[7] G.R. Liu. Neurons-samples theorem. *International Journal of Computational Methods*, 19(10):2250043, 2022.

[8] GR Liu. Solution existence theory for artificial neural networks. *International Journal of Computational Methods*, 19(09):2250040, 2022.

[9] G.R. Liu. *Machine Learning with Python: Theory and Applications.* World Scientific, New Jersey, 2023.

[10] G.R. Liu. *Numbers and Functions: Theory, Formulation, and Python codes.* World Scientific, New Jersey, 2024.

[11] G.R. Liu. *Machine Learning: Datasets for Classification with Python.* ScienTech Publisher, Mason, USA, 2025.

[12] G.R. Liu. *PINN with Python: An Introduction.* ScienTech Publisher, Mason, USA, 2025.

[13] G.R. Liu. *What is Machine Learning: An Introduction for Beginners.* ScienTech Publisher, Mason, USA, 2025.

[14] G.R. Liu. *Linear Algebra with Python: Modern Computational Approaches.* ScienTech Publisher, Mason, USA, 2026.

[15] G.R. Liu. *Machine Learning: Datasets for Regression with Python.* ScienTech Publisher, Mason, USA, 2026.

[16] G.R. Liu and X. Han. *Computational inverse techniques in nondestructive evaluation.* CRC Press, Taylor and Francis Group, New York, 2003.

[17] G.R. Liu and T.T. Nguyen. *Smoothed finite element methods.* CRC Press, New York, 2010.

[18] G.R. Liu and Siu Sin Quek. *The finite element method: a practical course, 2nd Ed.* Butterworth-Heinemann, London, 2013.

[19] Maziar Raissi, Paris Perdikaris, and George Em Karniadakis. Physics-informed neural networks: A deep learning framework for solving forward and inverse problems involving nonlinear partial differential equations. *Journal of Computational Physics*, 378:686–707, 2019.

[20] Walter Rudin. *Principles of Mathematical Analysis.* McGraw-Hill, New York, 3 edition, 1976.

[21] Joshua B. Tenenbaum, Vin De Silva, and John C. Langford. A global geometric framework for nonlinear dimensionality reduction. *Science*, 290(5500):2319–2323, 2000.

[22] Wikipedia contributors. Affine transformation — Wikipedia, the free encyclopedia, 2025. [Online; accessed 02-January-2026].

Index

Made in the USA
Las Vegas, NV
03 November 2026

57421907R00061